SpringerBriefs in Molecular Science

More information about this series at http://www.springer.com/series/8898

Sahar Amiri · Mohammad Ali Semsarzadeh
Sanam Amiri

Silicon Containing Copolymers

 Springer

Sahar Amiri
Chemical Engineering
Tabiat Modares University
Tehran
Iran

Sanam Amiri
Amir kabir University of Technology
Tehran
Iran

Mohammad Ali Semsarzadeh
Tabiat Modares University
Tehran
Iran

ISSN 2191-5407   ISSN 2191-5415   (electronic)
ISBN 978-3-319-09224-9   ISBN 978-3-319-09225-6   (eBook)
DOI 10.1007/978-3-319-09225-6

Library of Congress Control Number: 2014946375

Springer Cham Heidelberg New York Dordrecht London

Printed on acid-free paper

Springer is part of Springer Science+Business Media (www.springer.com)

Preface

The present volume examines the coherent scientific community devoted to the study of Silicon Containing Copolymers. It starts with a series of studies of silicone and silicon-based block copolymers that were important in the synthesis of new polymeric materials with new properties.

Recent advances in controlled radical polymerization techniques have led to facile synthesis of well-defined block copolymers with a wide range of functional monomers. The excellent properties of silicone polymers include high stabilities toward heat and ultraviolet radiation, very low glass transition and melting temperatures, and good gas permeability and low surface tension, and most importantly, polysiloxanes are nontoxic and environmentally compatible. The association between polysiloxanes and various polymers open the way to various industrial applications. Block and graft copolymers were prepared using ATRP. This volume describes the synthesis of novel triblock and pentablock copolymers based on PDMS macroinitiators with various vinyl monomers.

A history of the Amiri Discussions that led up to this event is presented. The book concludes with an essay on the history of silicone-based polymers' synthesis and characterization. The factors that influenced this history form a fascinating study of new findings in synthesis of various Poly(dimethyl siloxane)-based copolymers via atom transfer radical polymerization (ATRP) or cobalt-mediated radical polymerization (CMRP), the formation of a now thriving scientific research community.

Sahar Amiri

Acknowledgments

My interest in polymer and synthesis science was kindled by Mohammad A. Semsarzadeh as an undergraduate and fanned into flame while earning my Ph.D. in his laboratory at Tarbiat Modares University. He set a high standard both for science and for writing.

The basis of life as a scientist is a warm and supportive home life. My mother and father supported me in my whole life. It would have been impossible without them.

Contents

Chapter 1
Introduction

Abstract Recent advancements in controlled radical polymerization techniques
have led to the facile synthesis of the well-defined silicon-based thermoreversible
block copolymers. Because of their inorganic–organic structure and the flexibility
of the silicone bonds, silicones have some unique properties including thermal
oxidative stability, low temperature flow ability, high compressibility, low surface
tension, hydrophobicity, good electric properties and low fire hazard. Well-defined
silicon-based block copolymers were synthesized via ATRP of St, MMA, MA,
BDMP or VAc monomers at 60 °C using CuCl/PMDETA as a catalyst system
initiated from Br-PDMS-Br. Also new polyrotaxanes were synthesized based on
new copolymers via inclusion complex of γ-cyclodextrin and block copolymers.

Keywords Poly(dimethyl siloxane) (PDMS) · Atom transfer radical polymerization (ATRP) · Cobaltmediated controlled radical polymerization (CMCRP) ·
γ-Cyclodextrin · 4-Bromo-2,6-dimethylphenol (BDMP)

Silicones are the most versatile polymer known, are organo-metallic polymers
derived from sand which is the abundant raw material on earth. The majority of
silicone based polymers is poly(dimethyl siloxanes). Because of their inorganic-
organic structure and the flexibility of the silicone bonds, silicones have some unique
properties including thermal oxidative stability, low temperature flow ability, high
compressibility, low surface tension, hydrophobicity, good electric properties and
low fire hazard. These current special properties have encouraged us to explore
alternative method in synthesis of well-defined controlled microstructures of silicone
copolymers. Copolymers containing poly(dimethyl siloxane) (PDMS) have received
considerable attention due to their unique properties, such as very low glass transition temperature, low surface energy, low solubility parameter and physiological
inertness. Some of their specialty applications are in the fields of biomaterials and
surfactants. Though a variety of synthetic routes have been used to prepare PDMS
containing copolymers, less work has been done preparing block copolymers of
dimethyl siloxane with various vinyl monomers. Recent advancements in controlled
radical polymerization techniques have led to facile synthesis of the well-defined
silicon based block copolymers. PDMS block and graft copolymers are expected to

possess unique physical properties making them viable candidates for materials such as thermoplastic elastomers, compatibilizers for silicone rubbers, surfactants for supercritical carbon dioxide and thermoreversible block copolymers. Several synthetic methods for PDMS copolymers have been reported. The best results were obtained using living anionic polymerization. Atom transfer radical polymerization (ATRP) and cobalt-mediated controlled radical polymerization (CMCRP) are considered to synthesis new PDMS based homo and block copolymers with new properties. CMCRP particularly have a number of advantages over the already well-known oxidative coupling or phase transfer reactions reported earlier [1–2]. ATRP is initiated by alkyl halides, and therefore, any polymer that has a sufficiently active alkyl halide end-group could initiate ATRP to afford block copolymers. The purpose of this study was to expand the scope of the ATRP method to produce block copolymers from PDMS macroinitiators. It has been successfully employed in the synthesis of a large range of previously unknown well-defined block copolymers. An important advantage of CMCRP is the copolymerization reaction of various monomers such as vinyl acetate (VAc), styrene (St), poly(dimethyl siloxane) (PDMS), methyl methacrylate (MMA), methyl methacrylate (MA) and 4-bromo-2,6-dimethylphenol (BDMP). New architectural designs for controlled compositions of block copolymers is of a great need in the area of macromolecular design of membranes, foams, and drug delivery systems.

   In this study, inclusion complexes of γ-cyclodextrin (γ-CD) and boromoalkyl-terminated poly(dimethylsiloxane) (Br-PDMS-Br) were synthesized at room temperature for 7 days without utilizing sonic energy, for the first time and characterized with various methods. This inclusion complex could be used as macroinitiator in ATRP reactions which let us synthesis of new block copolymers of silicone suitable for membrane applications with well-define microstructure and favorite thermal properties.

   Well-defined silicon based block copolymers were synthesized via ATRP of St, MMA, MA, BDMP or VAc monomers at 60 °C using CuCl/N,N,N′,N″,N″-penta-methyldiethylenetriamine (PMDETA) as a catalyst system initiated from Br-PDMS-Br. Well-defined poly(vinyl acetate-b-dimethyl siloxane-b-vinyl acetate) (PVAc-b-PDMS-b-PVAc), poly(phenylene oxide-b-dimethyl siloxane-b-phenylene oxide) (PPO-b-PDMS-b-PPO) triblock and poly(vinyl acetate-b-styrene-b-dimethyl siloxane-b-styrene-b-vinyl acetate) (PVAc-b-PSt-b-PDMS-b-PSt-b-PVAc), Poly(vinyl acetate-b-styrene-b-dimethyl siloxane-b-styrene-b-vinyl acetate) (PVAc-b-PSt-b-PDMS-b-PSt-b-PVAc) and poly(methyl methacrylate-b-vinyl acetate-b-dimethyl siloxane- b-vinyl acetate-b-methyl methacrylate) (PMMA-b-PVAc-b-PDMS-b-PVAc-b-PMMA) pentablock copolymers were confirmed by ^{1}H-NMR, GPC and DSC.

   Recent advancements in controlled radical polymerization techniques have led to the facile synthesis of the well-defined silicon based thermoreversible block copolymers. ATRP has been utilized to develop well-defined functional thermoreversible block copolymers. Polyrotaxanes have attracted much attention in the past decades for their great potential as stimulus-responsive materials which can produce PDMS based thermoreversible block copolymers. In this study thermore-versible block copolymers were synthesized via 2 methods:

(1) Used inclusion complex between γ-CD and Br-PDMS-Br as macroinitiator in ATRP of various monomers
(2) Synthesis of pentablock copolymers and reacted with γ-CD in a CMRP

Polyrotaxane based on block copolymers can undergo a temperature-induced reversible transition upon heating of the copolymer complex from a white complex at 22 °C to a green complex in 55 °C which characterized with XRD and ^{1}H-NMR. XRD showed a change in crystallinity percent of St peak with changing the temperature.

Chapter 2
Polyrotaxane Based on Inclusion Complexes of OH-PDMS-OH and Br-PDMS-Br with γ-Cyclodextrin Without Utilizing Sonic Energy

Abstract γ-Cyclodextrin (γ-CD) was found to form inclusion complexes with poly(dimethylsiloxane)s (PDMS) under sonic energy and the products were crystalline compounds in high yields, which have been investigated extensively in the past. In this study, an inclusion complex between PDMS and γ-CD was synthesized at room temperature in the presence of light and mixing, in the absence of light and in the absence of mixing. These inclusion complexes (ICs) were characterized by XRD, DSC, ^{1}H-NMR and FT-IR spectroscopy. The findings suggest that the reaction conditions change the crystalline structure and mole ratios of the complexes (monomer unit/γ-CD) determined by ^{1}H-NMR spectroscopy for all of the ICs with γ-CD.

Keywords Poly(dimethylsiloxane) (PDMS) · γ-Cyclodextrin(γ-CD) · Inclusion complex formation · Sonic energy

2.1 Introduction

γ-cyclodextrin (γ-CD) formed inclusion complexes with bis(hydroxyalkyl)-terminated poly(dimethyl siloxane) (OH-PDMS-OH) and bis(2-bromoisobutyrate)-terminated poly(dimethyl siloxane) (Br-PDMS-Br) at room temperature for 7 days without utilizing sonic energy which gives crystalline compounds. Complexes of CDs with silicon-containing polymers are formed as new organic-inorganic hybrids with exact stoichiometric relationships. Cyclodextrins are cyclic oligosaccharides the most common consisting of 6, 7 and 8 glucopyranose units linked by R-1,4 glucosidic bonds; they are called α-cyclodextrin (α-CD), β-cyclodextrin (β-CD) and γ-cyclodextrin (γ-CD), respectively [1–3]. They present a truncated cone shape with a hydrophobic core, which can accommodate nonpolar compounds [4, 5] and two hydrophilic rims composed of -OH groups. The inclusion complex (IC) formation depends on internal parameters (the nature of the CD, polymer and solvent media) as well as on external parameters (temperature and pressure). Accordingly, new

© The Author(s) 2014
S. Amiri et al., *Silicon Containing Copolymers*,
SpringerBriefs in Molecular Science, DOI 10.1007/978-3-319-09225-6_2

strategies were developed to fabricate novel supramolecular hydrogels via several routes. Recently, attention has been paid on ICs formed by CDs and inorganic polymers which offer different sites of binding and may be selectively threaded by CDs [4–8]. In 1990, Harada and Kamachi discovered an inclusion complexation of many α-CDs and poly(ethylene glycol) (PEG) that resulted in the formation of polypseudorotaxanes [6]. This chapter describes the preparation and characterization of inclusion complexes of CDs with OH-PDMS-OH and Br-PDMS-Br at room temperature for 7 days.

The results indicated the successfully formation of an inclusion complex between γ-CD/PDMS and γ-CD/Br-PDMS-Br. Differential scanning calorimetry (DSC) analysis confirmed the existence of the complex with an endothermic melting peak of γ-CD at about 110 °C and the one at 310 °C disappearing [6–8].

2.2 Inclusion Complex of γ-CD and OH-PDMS-OH and Br-PDMS-Br

^{1}H-NMR of the functionalized OH-PDMS-OH was used to verify the quantitative modification of the end groups to the terminal bromine atoms (Fig. 2.1). The volatiles were removed under vacuum from the final product. GPC analysis showed $M_n = 14,800$, with a MWD = 1.93, maximum yield determined gravimetrically is 82 %. ^{1}H-NMR (CDCl$_3$) indicated $\delta = 0.0{-}0.3$ ppm for protons of methyl groups of -Si(CH$_3$)$_2$O, $\delta = 2.0$ ppm for methylene group next to the bromide [6, 9–11] (Scheme 2.1).

When OH-PDMS-OH or Br-PDMS-Br (liquid) was added to aqueous solutions of γ-CD (diameter of the cavity: 7.0 Å) and the mixture was mixed at room temperature for 7 days, the heterogeneous solution became turbid and the complexes were formed as crystalline precipitates. This is the first observation in which γ-CD forms a complex with inorganic polymers at room temperature without sonic energy. Table 2.1 shows the results of the complex formation between γ-CD and OH-PDMS-OH or Br-PDMS-Br at room temperature for 7 days. γ-CD forms a complex as soon as OH-PDMS-OH or Br-PDMS-Br is added and the reaction yield becomes higher after 6 days at the room temperature without sonic energy. The yield of γ-CD/OH-PDMS-OH and γ-CD/Br-PDMS-Br is 79 and 71 % respectively. The cavity of the γ-CD is large enough to accommodate PDMS [7, 12–14].

2.3 Stoichiometries

The complex formation of γ-CD with OH-PDMS-OH or Br-PDMS-Br was studied quantitatively. The amount of the complex formed increased with an increase in the amount of OH-PDMS-OH or Br-PDMS-Br added to the aqueous solution of γ-CD.

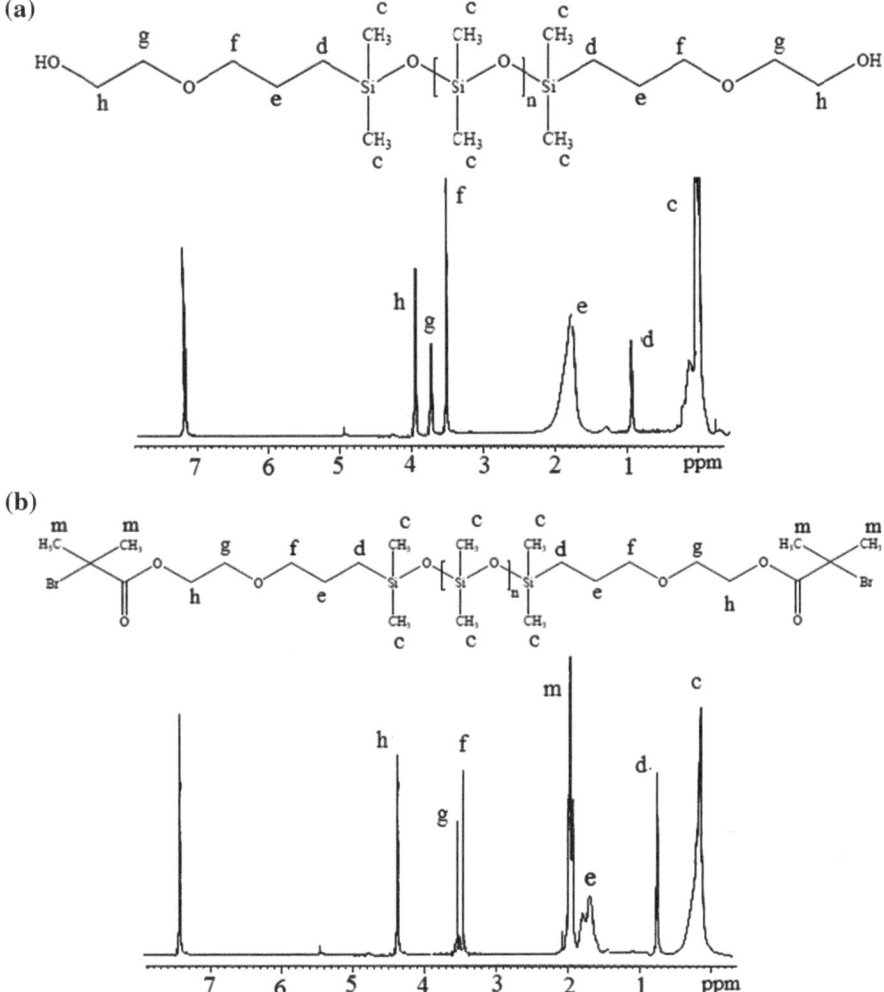

Fig. 2.1 ¹H-NMR spectra of **a** HO-PDMS-OH and **b** Br-PDMS-Br macroinitiator in the CDCl₃ solvent

The amount of the complex showed similar values even if excess amounts of OH-PDMS-OH or Br-PDMS-Br were used; this indicates the stoichiometric complexation. The continuous variation plot for the formation of the complex between γ-CD and OH-PDMS-OH or Br-PDMS-Br is at maximum level in 2:3 and 1:1 stoichiometry (monomer unit: γ-CD) respectively (Fig. 2.2). The stoichiometry was confirmed by the use of ¹H-NMR spectroscopy. The length of the 1.5 monomer units corresponds to the depth of the γ-CD cavity.

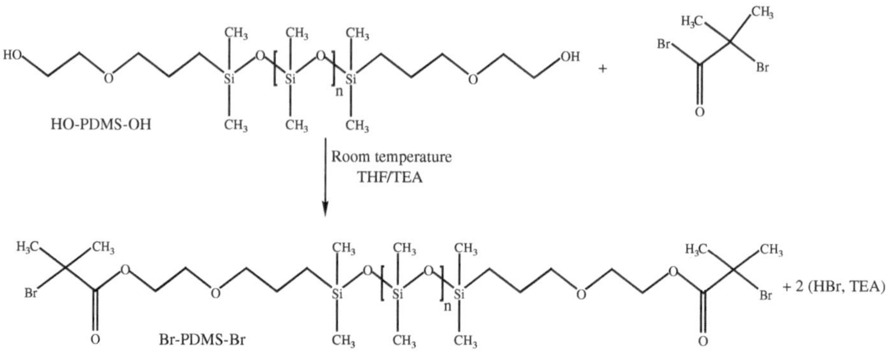

Scheme 2.1 Reaction scheme for the synthesis of bis(2-bromoisobutyrate)-terminated PDMS macroinitiator from bis(hydroxyalkyl)-terminated PDMS

Table 2.1 Assignments of FTIR of PDMS/γ-CD

Wavenumber (cm^{-1})	Assignment
2960	v(C-H) in CH$_3$
1259.90	δ(C-H) in Si-CH$_3$
791	v_a(Si-O-Si) in Si-O-Si
723	v_s(Si-O-Si) in Si-O-Si
601	P(C-H) in Si-CH$_3$
3369	v (O-H) in γ-CD

v = stretching mode, v_a = asymmetric stretching, v_s = symmetric stretching, δ = in-plane bending or scissoring, ρ = in-plane bending or rocking

2.4 Characterization Inclusion Complex Formation

The complexes were isolated by centrifugation, then washed, and dried. The inclusion complexes were thermally stable. The complexes were insoluble in water, even under boiling conditions [12, 15]. The FTIR spectrum of the PDMS/γ-CD is provided in Fig. 2.3. The spectrum showed strong Si-O-Si stretching absorptions at 400–800 cm^{-1}, which is characteristic of a siloxane backbone. The complete list of FTIR structure assignments is given in Table 2.1.

The X-ray diffraction pattern of the complex between γ-CD/OH-PDMS-OH and γ-CD/Br-PDMS-Br complexes at room temperature for 7 days shows that the complexes are crystalline (Fig. 2.4). The X-ray diffraction studies (powder) show that all of the complexes are crystalline, although linear PDMS is a liquid at room temperature. Harada reported that the crystal structures of CD complexes are classified mainly into three types: channel-type, cage-type, and layer-type [6].

The complexes were found to have a cage-type structure at room temperature (Fig. 2.4a, b). The reflection peaks γ-CD/OH-PDMS-OH and γ-CD/Br-PDMS-Br complexes are different from those of the γ-CD [12]. Figure 2.5 shows the ^{1}H-NMR

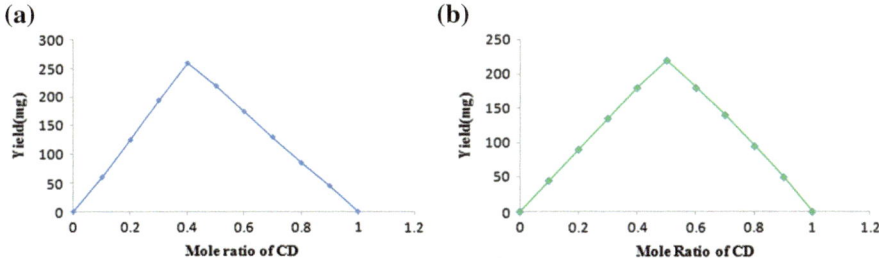

Fig. 2.2 Continuous variation plot for complex formation between γ-CD and **a** OH-PDMS-OH and **b** Br-PDMS-Br at room temperature for 7 days

(a)

(b)

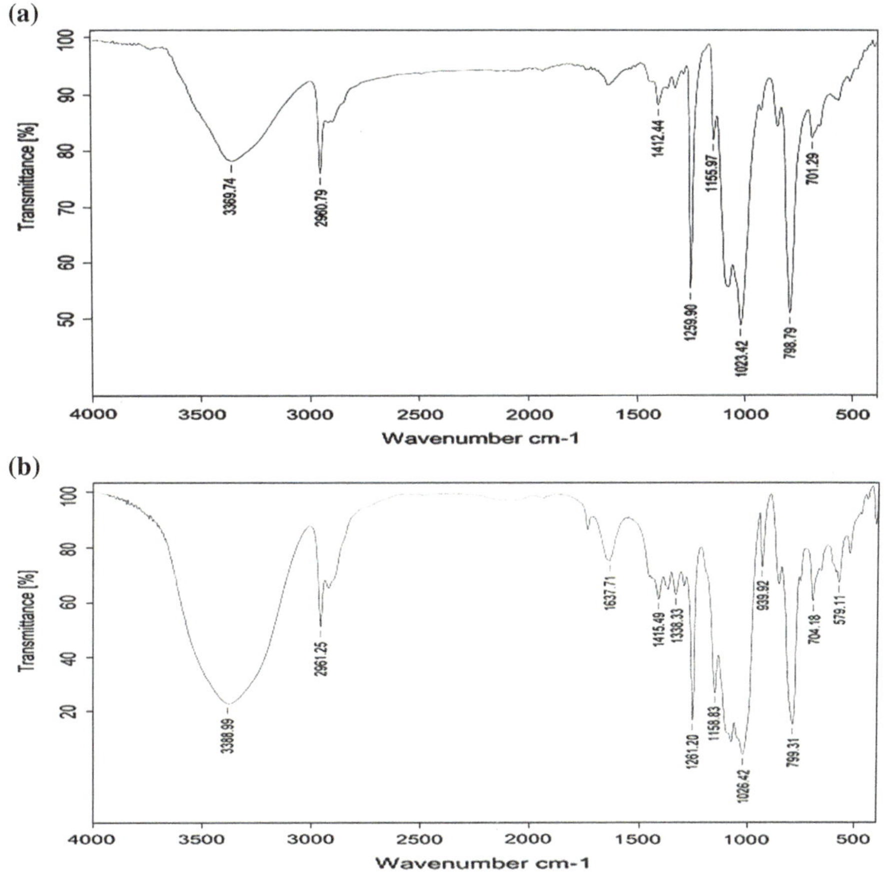

Fig. 2.3 FTIR of complex formation between γ-CD and (**a**) OH-PDMS-OH and (**b**) Br-PDMS-Br at room temperature for 7 days

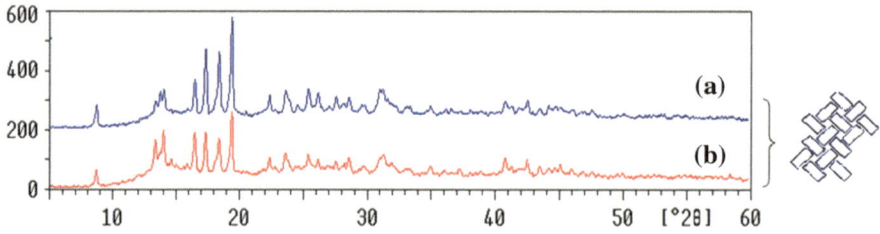

Fig. 2.4 XRD of the complex formation between γ-CD and **a** OH-PDMS-OH and **b** Br-PDMS-Br at room temperature for 7 days

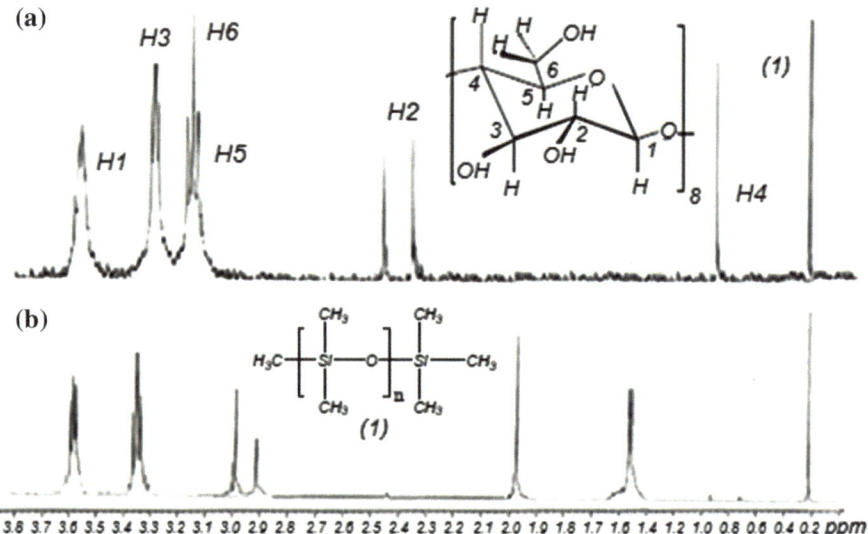

Fig. 2.5 ¹H-NMR spectrum of the complex between **a** OH-PDMS-OH/γ-CD 7 and **b** Br-PDMS-Br/γ-CD 7 day at room temperature

spectrum of the complex between γ-CD and PDMS. We calculated the mole ratio of OH-PDMS-OH or Br-PDMS-Br to γ-CD in the complexes. The moles ratios of the complexes are 2:3 and 1:1 (monomer unit: γ-CD) for OH-PDMS-OH/γ-CD and Br-PDMS-Br/γ-CD respectively, which is similar to those obtained from the conversion of complex (Fig. 2.1). The length of 2 monomer units corresponds to the depth of the γ-CD cavity.

DSC thermograms of γ-CD and the inclusion complex of OH-PDMS-OH/γ-CD and Br-PDMS-Br/γ-CD in the temperature range from −100 to +350 °C are shown in Fig. 2.6. DSC has been used to characterize the thermal and structural properties of many compounds. DSC is a useful tool to determine the melting and crystallization temperatures, which can provide both quantitative and qualitative information about

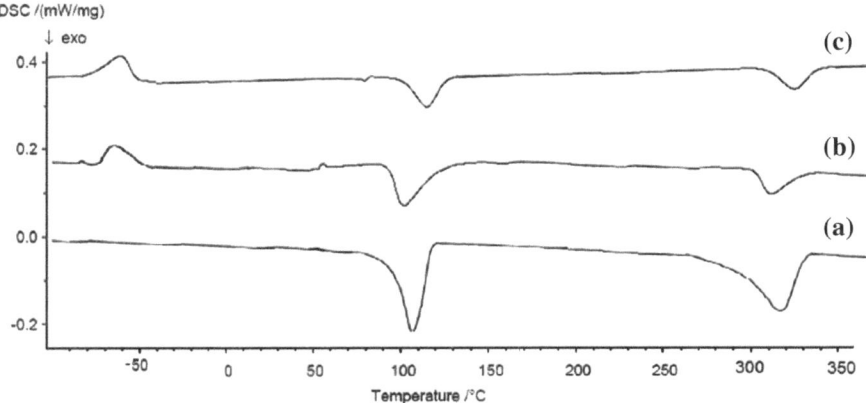

Fig. 2.6 DSC spectrum of **a** γ-CD, **b** the complex between OH-PDMS-OH/γ-CD and **c** Br-PDMS-Br/γ-CD 7 days at room temperature

the physiochemical state of the guest inside the CD complexes. The DSC plot of γ-CD, showed that two endothermic peaks were observed in the temperature range between 100 and 110 °C due to the loss of water and near 320 °C due to the γ-CD fusion [16, 17]. Glass transition temperature (T_g) of PDMS cannot be observed in Fig. 2.6, because its value has been reported to be about −120 °C [18–20]. The DSC thermograms for the OH-PDMS-OH/γ-CD and Br-PDMS-Br/γ-CD systems show sifting the persistence exothermic peak of PDMS (at −120 °C) in all products. These results are characteristic of the microphase separated morphology of the inclusion complex. On the basis of DSC results, one can conclude that the inclusion complex of OH-PDMS-OH/γ-CD and Br-PDMS-Br/γ-CD has successfully been synthesized.

2.5 Summary

This is the first observation that cyclodextrins have formed a complex with inorganic polymers at room temperature without sonic energy. These kinds of complexes may provide a new way of creating new organic-inorganic hybrids and other functional supramolecular architectures, especially inclusion complexes of γ-CD and OH-PDMS-OH or Br-PDMS-Br. The results showed that monomer structure may affect the monomer unit/γ-CD calculated through ^{1}H-NMR. The DSC results indicated that the inclusion complexes of OH-PDMS-OH/γ-CD and Br-PDMS-Br/γ-CD have successfully been synthesized.

References

1. Yang C, Yang J, Ni X, Li J (2009) Macromolecules 42:3856–3859
2. Sauvage J-P, Dietrich B (1999) Nature 401:150–152
3. Semlyen JA (2000) Cyclic polymers, 2nd edn. Kluwer Academic Publisher, New York
4. Bender ML, Komiyama M (1978) Cyclodextrin Chemistry. Springer, New York 99:5146–5151
5. Wenz G, Han BH, Muller A (2006) Chem Rev 106:782–817
6. Harada A, Kamachi M (1990) Macromolecules 23:2821–2823
7. Jones RG (2001) Applied Organometallic Chemistry Special Issue. Nanomaterials 15:440–441
8. Duo Q, Wang C, Cheng C, Han W, Thune PC, Ming W (2006) Macromol Chem Phys 207:2170
9. Gelsest C (1967) Reactive silicones: forging new polymer links. Gelest Inc., Morrisville, Pennsylvania 19067. Available at www.gelest.com/company/pdfs/reactivesilicones.pdf
10. Semsarzadeh MA, Abdollahi M (2011) J Appl Polym Sci 4:2423–2430
11. Miller RD, Michl J (1989) J Chem Rev 89:1359–1410
12. Trefonas P, Djurovich PI, Zhang X-H, West R, Miller RD, Hofer D (1983) J Polym Sci, Polym Lett Ed 21:819–823
13. Okumura H, Okad M, Kawaguchi Y, Harada A (2003) Macromolecules 36:6422–6429
14. Araki J, Zhao C, Ito K (2005) Macromolecules 38:7524–7527
15. Huh KM, Ooya T, Lee WK, Sasaki S, Kwon IC, Jeong SY, Yui N (2001) Macromolecules 34:8657–8662
16. Silver Stain RM, Bassler GC, Morrill TC (1981) Spectrometric Identification of Organic Compounds, 4th Ed. Wiley, New York, pp 166–169
17. Farcasa A, Jarrouxb N, Farcasc A-M, Harabagiua V, GueganbDigest P (2006) J Nanomaterials Biostructures 1(2):55–60
18. Clarson SJ, Dodgson K, Semlyen JA (1985) Polymer 26:930–934
19. Clarson SJ, Mark JE, Dodgson K (1988) Polym Commun 29:208–211
20. Luo ZH, Yu HJ, He TY (2008) J Appl Polym Sci 108:1201–1208

Chapter 3
Synthesis and Characterization of PDMS Based Triblock and Pentablock Copolymers

Abstract Poly(dimethylsiloxane)-based triblock and pentablock copolymers have been synthesized via atom transfer radical polymerization (ATRP) of styrene (St) and vinyl acetate (VAc) telomer at 60 °C in the presence of CuCl/PMDETA as a catalyst system initiated with bis(bromoalkyl)- terminated PDMS macroinitiator (Br-PDMS-Br). Vinyl acetate telomers prepared from radical and controlled radical telomerization with Co(acac)$_2$/DMF catalyst and ligand were used in atom transfer radical polymerization to synthesize poly(vinylacetate-b-dimethylsiloxane-b-vinylcetate) (PVAc-b-PDMS-b-PVAc) triblock copolymer and poly(vinyl acetate-b-styrene-b-dimethyl siloxane-b-styrene-b-vinyl acetate) (PVAc-b-PSt-b-PDMS-b-PSt-b-PVAc) pentablock copolymers. The PDMS-based triblock and pentablock copolymers revealed a significant effect of Co(acac)$_2$/DMF on PVAc telomere, which was used in the synthesis of highly ordered block copolymer on a well-defined microstructure. The results were confirmed by ^{1}H-NMR and DSC indicating that a low Tg of PDMS in the microstructure of block copolymer has made the block copolymer flexible for new applications.

Keywords Poly(vinyl acetate) (PVAc) · Poly(dimethylsiloxane) (PDMS) · Polystyrene (PSt) · Atom transfer radical polymerization (ATRP) · Block copolymers · Telomerization · Co(acac)$_2$

3.1 Introduction

Poly(dimethyl siloxane) (PDMS) based triblock and pentablock copolymers were synthesized via atom transfer radical polymerization (ATRP) of vinyl acetate (VAc), styrene (St), methyl acrylate (MA) and methyl methacrylate (MMA) monomers at 60 °C using CuCl/N,N,N′,N″,N″-pentamethyldiethylenetriamine (PMDETA) as a catalyst system initiated from boromoalkyl-terminated poly (dimethylsiloxane) (Br-PDMS-Br) macroinitiator. Well-defined poly(vinyl acetate-b-dimethyl siloxane-b-vinyl acetate) (PVAc-b-PDMS-b-PVAc) triblock and poly

© The Author(s) 2014

S. Amiri et al., *Silicon Containing Copolymers*,

SpringerBriefs in Molecular Science, DOI 10.1007/978-3-319-09225-6_3

(vinyl acetate-b-styrene-b-dimethyl siloxane-b-styrene-b-vinyl acetate) (PVAc-b-PSt-b-PDMS-b-PSt-b-PVAc), poly(vinyl acetate-b-styrene-b-dimethyl siloxane-b-styrene-b-vinyl acetate) (PVAc-b-PSt-b-PDMS-b-PSt-b-PVAc) and poly(methyl methacrylate-b-vinyl acetate-b-dimethyl siloxane-b-vinyl acetate-b-methyl methacrylate) (PMMA-b-PVAc-b-PDMS-b-PVAc-b-PMMA) pentablock copolymers were confirmed by ^{1}H-NMR, GPC and DSC.

Silicones have been used as co-monomers incorporating other monomers and polymers to yield new properties. Condensation polymerization has been widely utilized to prepare the copolymers of silicone with urethane and amides [1, 2]. The excellent properties of silicones with their marvelous thermal properties and stability, and relatively low glass transition temperatures have made this polymer ideal for manufacturing membranes [3–5]. Authors of the current study have addressed earlier the synthesis of block copolymers with silicones containing styrene (St) via atom transfer radical polymerization (ATRP) with CuCl/PMDETA catalyst [6, 7]. To make the new block copolymers of silicone suitable for membrane applications, silicone macroinitiator has been used in ATRP reaction with vinyl monomers, while highly rubbery silicone segments are designed in a macromolecular structure with a more rigid vinyl segments. This produces a new no-rigid block copolymer with a well-define microstructure and favorite thermal properties [8, 9]. In this study, thermal properties of the triblock and pentablock copolymers was investigated by DSC which revealed a nearly available way to synthesize polymers with soft segments of PDMS at $T_g \approx -120\ °C$ [6], PMMA at 89 °C, PSt at 110 °C and PVAc segments at 45–50 °C [7–10]. This is known as a suitable material to synthesize a rigid block of the new block copolymers from PDMS. These block copolymers provided a method to design the block copolymer with PDMS segment, allowing us for the adjustments in the flexibility by PDMS and large effect of softening point (T_g), or rigidity from styrene, methyl acrylate or methyl methacrylate monomers from the higher melting temperature (T_m).

3.2 Characterization of the PDMS Macroinitiator (Br-PDMS-Br)

^{1}H-NMR of the functionalized PDMS was used to verify the quantitative modification of the end groups to the terminal bromine atoms (Fig. 3.1). The volatiles were removed under vacuum from the final product. GPC analysis showed $M_n = 14,800$, with a MWD = 1.93, maximum yield determined gravimetrically is 82 %. ^{1}H-NMR (CDCl$_3$) indicated $\delta = 0.0–0.3$ ppm for protons of methyl groups of $-Si(CH_3)_2O$, $\delta = 2.0$ ppm for methylene group next to the bromide (Scheme 3.1) [6, 7].

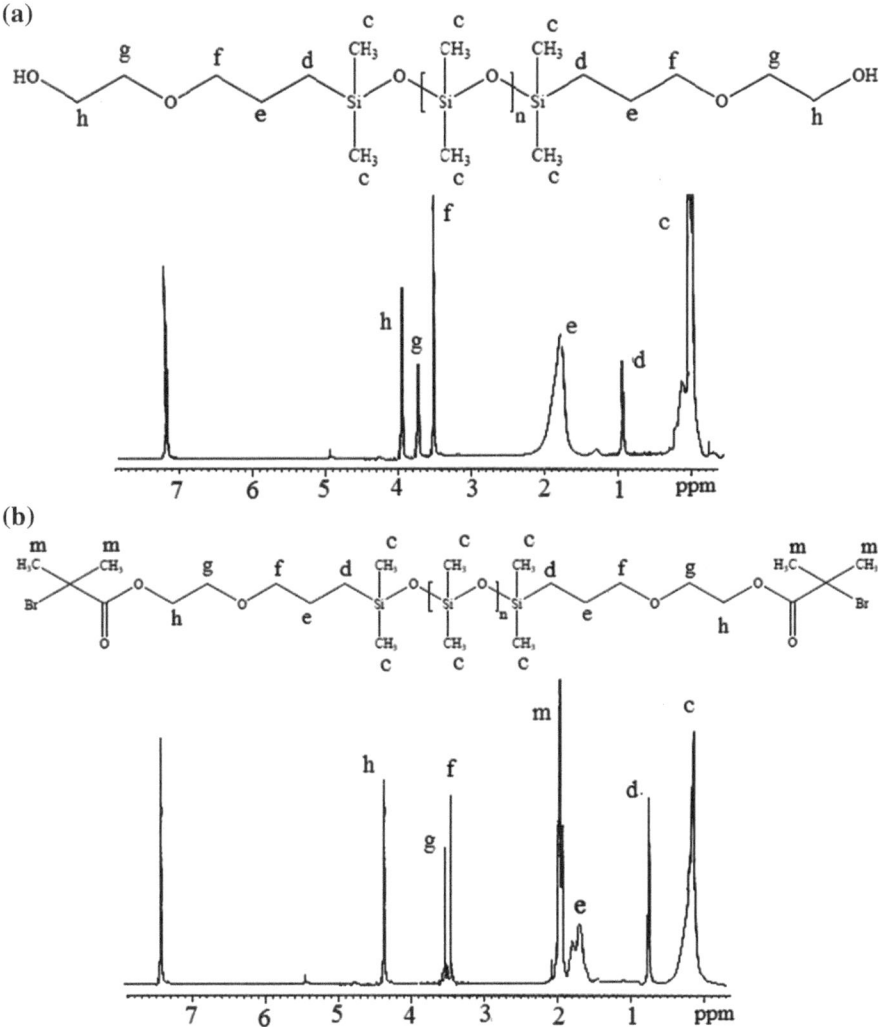

Fig. 3.1 ^{1}H-NMR spectra of HO-PDMS-OH **a** and Br-PDMS-Br macroinitiator **b** in the CDCl$_3$ solvent

3.3 PVAc-b-PDMS-b-PVAc Triblock Copolymers

First-order kinetic polymerization plots of PVAc initiated by Br-PDMS-Br mac-roinitiator and number average molecular weight, M_n, versus conversion in this polymerization are depicted in Fig. 3.2. The linear fit demonstrated that the constant concentration of propagating radical species and radical termination reactions are not significant. This was further supported by the linear increase in the molecular weight upon conversion, indicating that the number of chains remains constant

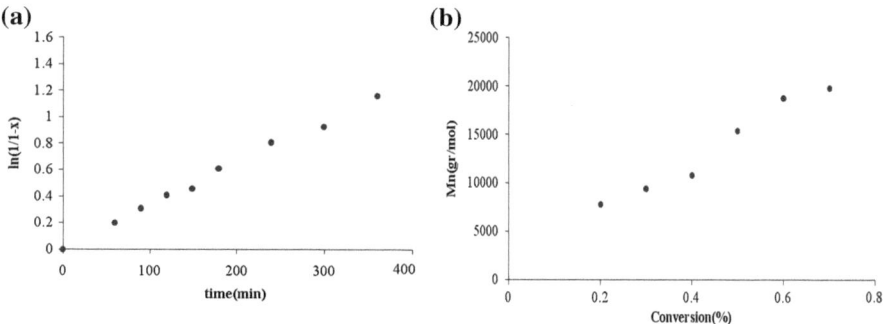

Scheme 3.1 Reaction scheme for the synthesis of bis(2-bromoisobutyrate)-terminated PDMS macroinitiator from bis(hydroxyalkyl)-terminated PDMS

(a)

y-axis: $\ln(1/1-x)$ (from 0 to 1.6)
x-axis: time(min) (from 0 to 400)

(b)

y-axis: Mn(gr/mol) (from 0 to 25000)
x-axis: Conversion(%) (from 0 to 0.8)

Fig. 3.2 **a** Time dependence of $\ln[M]_0/[M]$ (*M* monomer); **b** dependence of the polyvinyl acetate M_n on the monomer conversion for the ATRP of PVAc initiated with PDMS at 60 °C

during the reaction. The molecular weight distribution is also decreased with progress of the polymerization, implying that nearly all the chains have started to grow simultaneously (Fig. 3.2).

3.3.1 Characterization of PVAc-PDMS-b-PVAc Triblock Copolymer

The controlled polymerization uncovered a narrow molecular weight distribution which was in a good agreement with both theoretical and experimental molecular weights (Table 3.1). GPC results during the ATRP reaction of PVAc with PDMS macroinitiator are shown in Table 3.1.

Table 3.1 Summary of the results obtained from GPC and ^{1}H-NMR analysis for PVAc-PDMS-b-PVAc block copolymers at 60 °C for 6 h

(Co)Polymer	X (%)[a]	$M_{n,theory}$ (g mol^{-1})[b]	$M_{n,GPC}$ (g mol^{-1})	PDI
Br-PDMS-Br	87	14,282	14,800	1.93
PVAc-b-PDMS-b-PVAc	54	18,500	19,000	1.45

[a] Final conversion measured by gravimetric method

[b] $M_{n,theory} = 2\left(M_{n,Macroinitiator} + \frac{[M]_{PDMS}}{[M]_{Macroinitiator}} \times M_{n,PDMS} \times Conversion\right)$

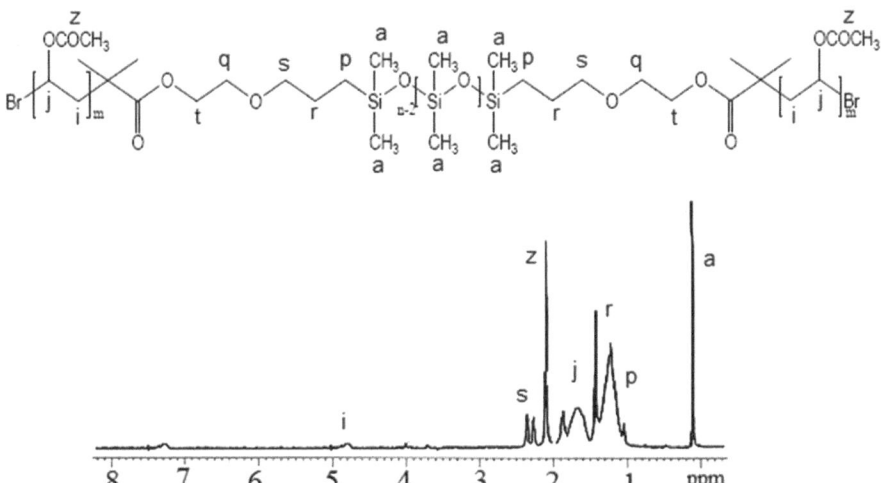

Fig. 3.3 ^{1}H-NMR spectrum of the ATRP of PVAc initiated with Br-PDMS-Br at 60 °C for 6 h

^{1}H-NMR spectra of the PVAc-b-PDMS-b-PVAc triblock copolymers have been illustrated in Fig. 3.3. All signals of the ^{1}H-NMR spectra were assigned to their corresponding monomers and it can be declared that the synthesis of triblock copolymers have been proceeded successfully.

The ^{1}H-NMR of triblock showed a signal at 0.0–0.2 ppm which was attributed to CH$_3$-Si methyl protons of PDMS, while the signals at 2.1–2.3 ppm correspond to CH$_3$ group of the PVAc segment [5, 10]. The area ratio of 6H signal from PSi to that of vinyl acetate showed that the ratio of sequence lengths of PVAc with respect to PDMS was found as X = (A/6)/(A*/6) = 2.14 (A* and A represent areas for the protons of methyl groups of –Si(CH$_3$)$_2$O and vinyl protons, respectively). It was possible to synthesize the PDMS-based triblock copolymers via ATRP from vinyl acetate in the presence of OH-PDMS-OH macroinitiator [10–12]. DSC thermograms of PVAc-b-PDMS-b-PVAc initiated with Br-PDMS-Br at 60 °C for PDMS-based triblock copolymers at the range of −150 to 150 °C are depicted in Fig. 3.4. PDMS and PVAc segments in the corresponding triblock copolymers exhibited a T$_g$ value of −120 °C for the soft block [6] as well as a 45–50 °C for the rigid block [10]. It can be

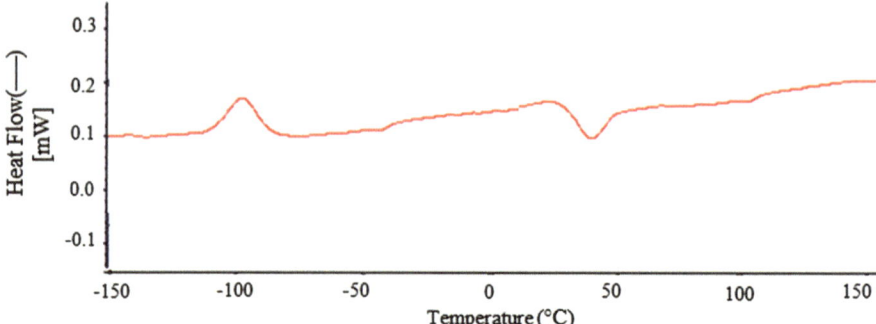

Fig. 3.4 DSC thermograms for the ATRP of PVAc initiated with Br-PDMS-Br from −150 to 150 °C with heating rate of 10 °C/min

inferred from Fig. 3.4 that the T_g values of PVAc segments are somewhat decreased using flexible PDMS segments in the center of the triblock copolymers [6].

3.4 PVAc-b-PSt-b-PDMS-b-PSt-b-PVAc Pentablock Copolymers

First-order kinetic polymerization plots of St and PVAc initiated by PDMS macroinitiator indicated that concentration of propagating radical species is constant and radical termination reactions are not significant on the time scale of the reaction. Within the bounds of experimental since the molecular weight distribution remained low during the polymerization, nearly all the chains were expected to have been started and grow simultaneously during an equilibrium step [26].

3.4.1 Characterization of PVAc-b-PSt-b-PDMS-b-PSt-b-PVAc Pentablock Copolymers

The results obtained from GPC are shown in Table 3.2 and there is good agreement between the theoretical and experimental molecular weight GPC results [26].

^{1}H-NMR spectra of the PVAc-b-PSt-b-PDMS-b-PSt-b-PVAc pentablock copolymer have been demonstrated in Fig. 3.5. All signals of the ^{1}H-NMR spectra were assigned to their corresponding monomers and it can be declared that the synthesis of pentablock copolymers have proceeded successfully [12, 13]. The ^{1}H-NMR of pentablock showed a signal at 0.0–0.2 ppm associated with CH$_3$-Si methyl protons of PDMS along with the signals at 2.1–2.3 and 6.6–7.1 ppm corresponded to OCOCH$_3$ group from the PVAc segment and PSt segment respectively. Therefore, it is possible to synthesize the PDMS-based pentablock

Table 3.2 Results obtained from GPC and ^{1}H-NMR Analyses for the PVAc-b-PSt-b-PDMS-b-PSt-b-PVAc pentablock copolymers

(Co)Polymer	X (%)[a]	$M_{n,theory}$ (g mol^{-1})[b]	$M_{n,GPC}$ (g mol^{-1})	PDI
Br-PDMS-Br	87	14,282	14,800	1.93
PVAc-b-PSt-b-PDMS-b-PSt-b-PVAc	54	20,300	21,700	1.5

[a] Final conversion measured by gravimetric method

[b] $M_{n,theory} = 2(M_{n,Macroinitiator} + \frac{[M]_{PDMS}}{[M]_{Macroinitiator}} \times M_{n,PDMS} \times Conversion)$

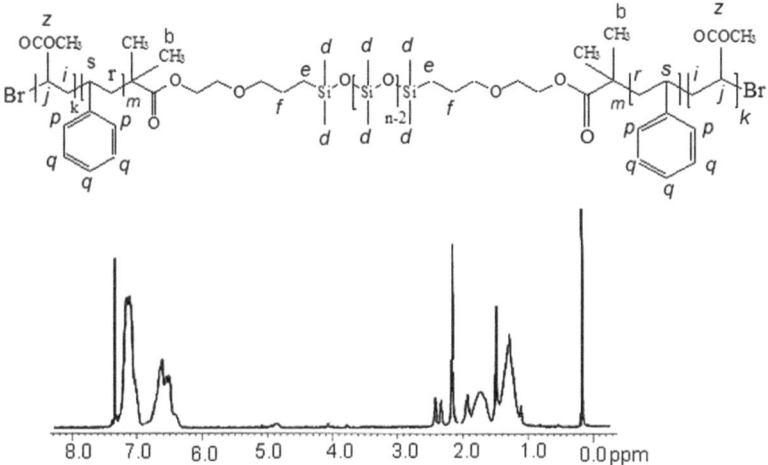

Fig. 3.5 ^{1}H-NMR traces for the PVAc-b-PSt-b-PDMS-b-PSt-b-PVAc pentablock copolymers initiated with Br-PDMS-Br and St and VAc as monomers at 60 °C for 6 h

copolymers via ATRP of monomers such as vinyl acetate and styrene in the presence of bis(haloalkyl)-terminated PDMS macroinitiator [12, 13].

The well-define microstructure can be used in macromolecular design of the pentablock copolymer, assuming one can increase the mole ratio of the monomer or used different molecular weight of the VAc or Br-PDMS-Br macroinitiator in the structure [26].

The ratio of sequence lengths of PVAc and PDMS was 2.17 and for PSt to PDMS was 2.38, so pentablock copolymers of PVAc-b-PSt-b-PDMS-b-PSt-b-PVAc were synthesized. Therefore, it is possible to synthesize the PDMS-based pentablock copolymers via ATRP of monomers such as vinyl acetate and styrene in the presence of OH-PDMS-OH macroinitiator [12, 13]. DSC thermograms of PDMS-based block copolymers are depicted in Fig. 3.6 for temperature range of −100 to +200 °C. Glass transition temperature (T_g) of PDMS is expected to be about −120 °C, thus it is observed in Fig. 3.6 at −59 °C [6, 10]. PVAc and PSt segments exhibited T_g values of 40 and 105 °C [9, 12, 13], respectively for the

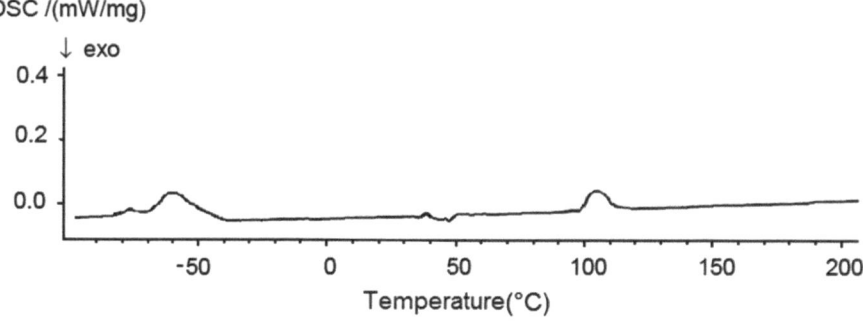

Fig. 3.6 DSC thermograms for the PVAc-b-PSt-b-PDMS-b-PSt-b-PVAc pentablock copolymers initiated with Br-PDMS-Br and St and VAc as monomers at 60 °C for 6 h

corresponding block copolymers. Considering the DSC results, one can conclude that the PDMS-based penta-block copolymers have been synthesized successfully.

3.5 PSt-b-PMMA-b-PVAc-b-PDMS-b-PVAc-b-PMMA-b-PSt and PSt-b-PMA-b-PVAc-b-PDMS-b-PVAc-b-PMA-b-PSt Block Copolymers

First-order kinetic polymerization plots of St, MA, MMA and PVAc initiated by Br-PDMS-Br macroinitiator are shown in Fig. 3.7.

The linear fit indicated that the concentration of propagating radical species is constant and radical termination reactions are not significant over time scale of the reaction. It was found that the molecular weights increase almost linearly with conversion, indicating that the number of chains was constant and the chain transfer reactions were rather negligible [16–18]. Thus from Fig. 3.7, one may conclude that the polymerization was controlled and had narrow molecular weight distribution.

3.5.1 Characterization of PSt-b-PMMA-b-PVAc-b-PDMS-b-PVAc-b-PMMA-b-PSt and PSt-b-PMA-b-PVAc-b-PDMS-b-PVAc-b-PMA-b-PSt Block Copolymers

GPC traces for the progress of ATRP reaction of PVAc, PSt, PMA and PMMA initiated with Br-PDMS-Br macroinitiator is represented in Fig. 3.8 (Table 3.3). The growing peak of block copolymers remained monomodal throughout the reaction, indicating that all the unreacted attachable initiators have been removed prior to polymerization [9, 13, 19–21].

(a)

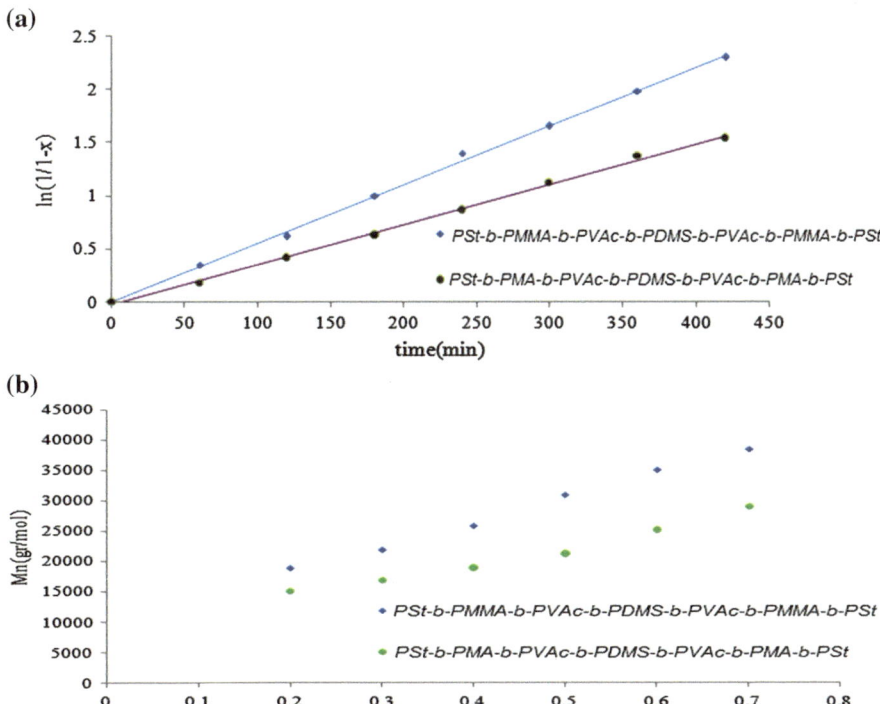

(b)

Fig. 3.7 **a** Time dependence of ln[M]$_0$/[M] (*M* monomer); **b** dependence of the block copolymer M$_n$ on the conversion for the ATRP of St, MA, MMA and VAc with Br-PDMS-Br macroinitiator at 60 °C

Fig. 3.8 GPC traces for the **a** PSt-b-PMMA-b-PVAc-b-PDMS-b-PVAc-b-PMMA-b-PSt and **b** PSt-b-PMA-b-PVAc-b-PDMS-b-PVAc-b-PMA-b-PSt block copolymers initiated with Br-PDMS-Br and MMA, MA,VAc and St as monomers at 60 °C for 6 h

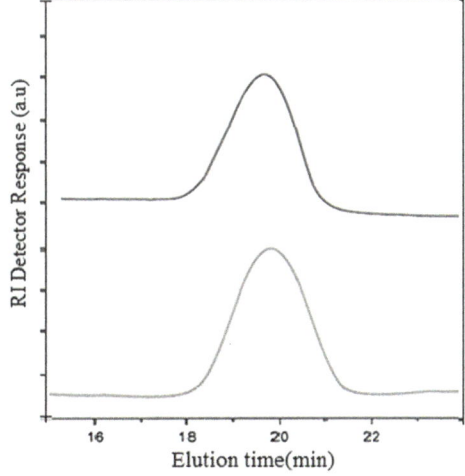

Table 3.3 Results obtained from GPC analysis for PSt-b-PMMA-b-PVAc-b-PDMS-b-PVAc-b-PMMA-b-PSt and PSt-b-PMA-b-PVAc-b-PDMS-b-PVAc-b-PMA-b-PSt block copolymers

(Co)Polymer	X (%)[a]	$M_{n,GPC}$ (gmol^{-1})	PDI
Br-PDMS-Br	87	14,800	1.93
PSt-b-PMMA-b-PVAc-b-PDMS-b-PVAc-b-PMMA-b-PSt	75	38,500	1.36
PSt-b-PMA-b-PVAc-b-PDMS-b-PVAc-b-PMA-b-PSt	54	28,900	1.45

[a] Final conversion measured by a gravimetric method

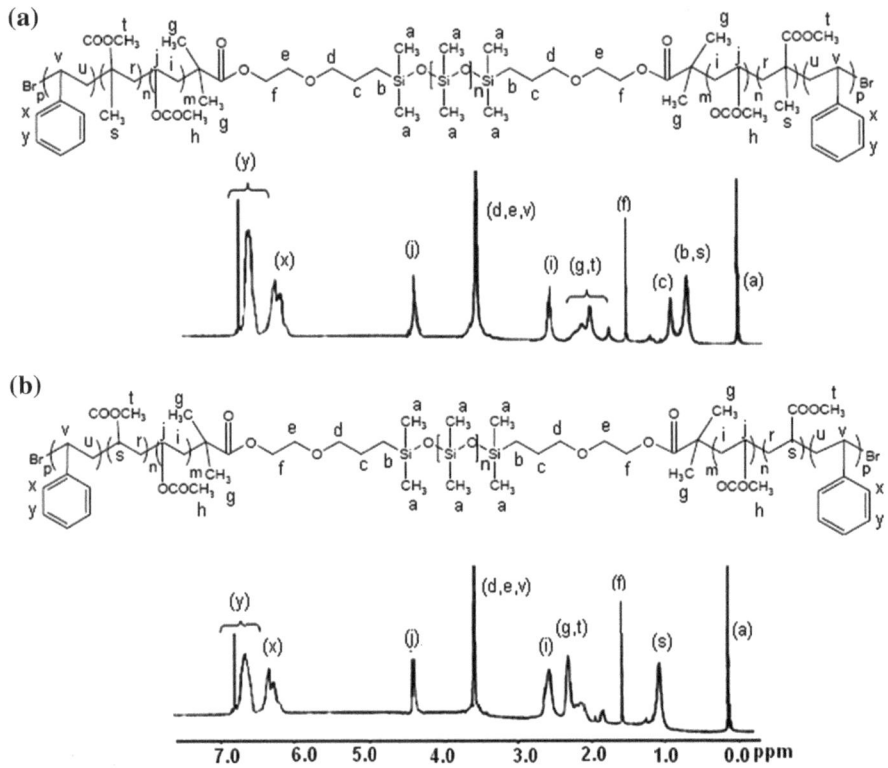

Fig. 3.9 ^{1}H-NMR traces for the PSt-b-PMMA-b-PVAc-b-PDMS-b-PVAc-b-PMMA-b-PSt and PSt-b-PMA-b-PVAc-b-PDMS-b-PVAc-b-PMA-b-PSt block copolymers initiated with PDMS macroinitiator and MMA, MA, St and VAc as monomers at 60 °C

^{1}H-NMR spectra of the PSt-b-PMMA-b-PVAc-b-PDMS-b-PVAc-b-PMMA-b-PSt and PSt-b-PMA-b-PVAc-b-PDMS-b-PVAc-b-PMA-b-PSt block copolymers have been demonstrated in Fig. 3.9. All signals of the ^{1}H-NMR spectra were assigned to their corresponding monomers and it can be declared that the synthesis of block copolymers have proceeded successfully [22–25]. The ^{1}H-NMR of block copolymers showed a signal at 0.0–0.2 ppm associated with CH$_3$-Si methyl protons of PDMS along with the signals at 2.1–2.3 and 6.6–7.1 ppm corresponded to

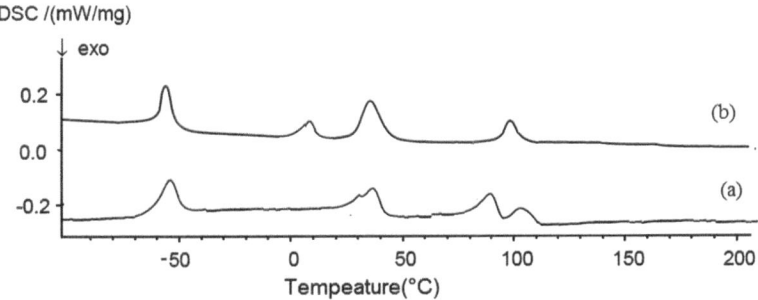

Fig. 3.10 DSC thermograms for the **a** PSt-b-PMMA-b-PVAc-b-PDMS-b-PVAc-b-PMMA-b-PSt and **b** PSt-b-PMA-b-PVAc-b-PDMS-b-PVAc-b-PMA-b-PSt block copolymers initiated with Br-PDMS-Br macroinitiator and MMA, MA,VAc and St as monomers at 60 °C for 6 h

OCOCH$_3$ group from the PVAc segment and PSt segment respectively. Signals at 0.8–1.0, 1.9–2.1 and 3.6 ppm corresponded to CH$_3$, CH$_2$ and COOCH$_3$ group from the PMMA and PMA segment [26–29]. Therefore, it is possible to synthesize the novel PDMS-based block copolymers via ATRP of monomers such as vinyl acetate, methyl methacrylate, methyl acrylate and styrene in the presence of bis (haloalkyl)-terminated PDMS macroinitiator [29, 30]. The well-define microstructure can be used in macromolecular design of the block copolymer, assuming one can increase the mole ratio of the monomer or used different molecular weight of the PDMS macroinitiator in the structure [26, 28].

The ratio of sequence lengths of PVAc and PDMS was 2.08, PMA and PDMS was 2.18, PSt and PDMS was 2.38 and for PMMA to PDMS was 2.34, so block copolymers of PSt-b-PMMA-b-PVAc-b-PDMS-b-PVAc-b-PMMA-b-PSt and PSt-b-PMA-b-PVAc-b-PDMS-b-PVAc-b-PMA-b-PSt were synthesized. Therefore, it is possible to synthesize the PDMS-based block copolymers via ATRP of monomers such as vinyl acetate, methyl acrylate, methyl methacrylate and styrene in the presence of bis(haloalkyl)-terminated PDMS macroinitiator [26–28].

DSC thermograms of PDMS-based block copolymers are depicted in Fig. 3.9 for temperature range of −100 to +200 °C. Glass transition temperature (T$_g$) of PDMS is expected to be about −120 °C, thus it is observed in Fig. 3.10 at −60 °C. PMMA, PMA, PVAc and PSt segments exhibited T$_g$ values of 90, 5, 40 and 105 °C [26–29], respectively for the corresponding block copolymers. Considering the DSC results, one can conclude that PDMS-based pentablock copolymers have been synthesized successfully.

3.6 Summary

Poly(dimethyl siloxane) (PDMS)-based triblock and pentablock copolymers were successfully synthesized using ATRP of PVAc, PSt, PMMA and PMA initiated with Br-PDMS-Br in the presence of CuCl/PMDETA as a catalyst system at 60 °C. Copolymers are of major interest because of the possibility of preparing amphiphilic pentablock copolymers by hydrolysis of the PVAc blocks. ^{1}H-NMR, DSC and GPC confirmed a well-defined triblock and pentablock copolymers synthesized. There was a very good agreement between the number-average molecular weight calculated from ^{1}H-NMR spectra and that calculated theoretically. The narrow dispersity indices from GPC of the synthesized pentablock copolymers yielded a lower PDI block copolymer indicating the living/controlled characteristic of the reaction.

References

1. Braunecker WA, Matyjaszewski K (2007) Prog Polym Sci 32:93–146
2. Chen M, Moad G, Rizzardo E (2011) Aust J Chem 64:433–437
3. Iovu MC, Matyjaszewski K (2003) Macromolecules 36:9346–9354
4. Harper CA, Petrie EA (2003) Plastics materials and processes: a concise encyclopedia. Wiley, Hoboken
5. Inoue H, Ueda A, Nagai S (1988) Polym Sci Part A Polym Chem 26:1077–1092
6. Matyjaszewski K, Nakagawa Y, Jasieczek CB (1998) Macromolecules 31:1535–1541
7. Duo Q, Wang C, Cheng C, Han W, Thune PC, Ming W (2006) Macromol Chem Phys 207: 2170–2179
8. Borkar S, Sen A (2005) J Polym Sci, Part A: Polym Chem 43:3728–3736
9. Semsarzadeh MA, Abdollahi M (2011) J Appl Polym Sci 4:2423–2430
10. Semsarzadeh MA, Abdollahi M (2009) J Appl Polym Sci 114:2509–2521
11. Semsarzadeh MA, Mirzaei A, Vasheghani-Farahani E, Nekoomanesh HM (2003) Eur Polym J 39:2193–2201
12. Semsarzadeh MA, Amiri S (2012) J Polym Res 19:9891–9900
13. Semsarzadeh MA, Amiri S (2012) J Chem Sci 2:521–527
14. David G, Boyer C, Tonnar J, Ameduri B, Lacroix-Desmazes P, Boutevin B (2006) Chem Rev 106:3936–3962
15. Tonnar J, Pouget E, Lacroix-Desmazes P, Boutevin B (2008) Eur Polym J 44:318–328
16. Koumura K, Satoh K, Kamigaito M, Okamoto Y (2006) Macromolecules 39:4054–4061
17. Pouget E, Tonnar J, Eloy C, Lacroix-Desmazes P, Boutevin B (2006) Macromolecules 39: 6009–6016
18. Lacroix-Desmazes P, Tonnar J, Boutevin B (2007) Macromol Symp 248:150–157
19. Shinoda H, Matyjaszewski K (2001) Macromol Rapid Commun 22:1176–1181
20. Wang TL, Liu YZ, Jeng BC, Cai YC (2005) J Polym Res 2:67–75
21. Peng H, Cheng S, Fan Z (2004) J Appl Polym Sci 92(6):3764–3770
22. Min K, Hu J, Wang CC (2002) J Polym Sci, Part A: Polym Chem 40:892–900
23. Ozturk T, Cakmak I (2008) J Polym Res 3:241–247
24. Huang CF, Kuo SW, Chen JK (2005) J Polym Res 6:449–456
25. Jakubowski W, Matyjaszewski K (2006) Angew Chem 27:4594–4598
26. Semsarzadeh MA, Amiri S (2013) J Inorg Organomet Polym 23:432–438
27. Semsarzadeh MA, Amiri S (2013) J Incl Phenom Macrocycl Chem. doi:10.1007/s10847-013-0330-1
28. Semsarzadeh MA, Amiri S (2013) J Inorg Organomet Polym 23:553–559
29. Semsarzadeh MA, Amiri S (2012) Silicon 4:151–156
30. Storsberg A, Helmut R (2000) Macromol Rapid Commun 21:1342–1346

Chapter 4
Cobalt Mediated Radical Polymerization of 4-Bromo-2,6-Dimethyl Phenol and Its Copolymerization with Poly(dimethyl siloxane) in the Presence of Co(acac)$_2$: DMF Catalyst

Abstract Novel well-defined poly(phenylene oxide) (PPO)-based block copolymers were polymerized from 4-bromo-2,6-dimethylphenol (BDMP) in the presence of a cobalt acetylacetonate (Co(acac)$_2$) catalyst with dimethyl formamide (DMF) as a ligand and initiated with benzoyl peroxide (BPO) at 60 °C in a novel controlled copolymerization reaction within a good yield. The monomer copolymerized with vinyl acetate (VAc), poly(methyl methacrylate)-b-poly(dimethyl siloxane)-b-poly (methyl methacrylate) (PMMA-b-PDMS-b- PMMA) and polystyrene-b-poly-(dimethyl siloxane)-b-polystyrene (PSt-b-PDMS-b-PSt) triblock copolymers in good yield. Characterization indicated a very narrow molecular weight distribution. The block copolymer indicated a new microstructure of phenylene oxide which was analyzed by proton-nuclear magnetic resonance (^{1}H-NMR), Fourier transform infrared spectroscopy (FTIR), differential scanning calorimetriy (DSC) and gel permeation chromatography (GPC). Meanwhile, the number average molecular weights calculated from the ^{1}H-NMR spectra were in very good agreement with the theoretically calculated values. The DSC results also indicated the successful formation of these new block copolymers.

Keyword Poly(dimethyl siloxane) (PDMS) · Poly(phenylene oxide)s(PPO) · 4-Bromo-2,6-dimethyl phenol (BDMP) · Co(acac)$_2$ catalyst · Controlled radical polymerization · Dimethyl formamide (DMF) · Reactivity ratio · Poly(methyl methacrylate) (PMMA)

4.1 Introduction

Cobalt-mediated controlled radical polymerization (CMCRP) is considered for the synthesis of new homopolymers and block copolymers with new properties. CMCRP particularly have a number of advantages over the already well-known oxidative coupling or phase transfer reactions reported earlier [1, 2]. An important advantage of CMCRP is copolymerization reaction of various monomers such as vinyl acetate,

© The Author(s) 2014
S. Amiri et al., *Silicon Containing Copolymers*,
SpringerBriefs in Molecular Science, DOI 10.1007/978-3-319-09225-6_4

styrene, poly(dimethyl siloxane) (PDMS), methyl methacrylate (MMA) and 4-bromo-2,6-dimethylphenol (BDMP). New architectural designs for controlled compositions of aromatic ethers is of a great need in the area of macromolecular design of membranes, foams, and drug delivery systems [3, 4]. Poly(phenylene oxide) (PPO) was synthesized from 4-bromo-2,6-dimethylphenol (BDMP) in the presence of cobalt acetoacetonate (Co(acac)$_2$) catalyst with dimethyl formamaide (DMF) ligand. The reaction was initiated with benzoyl peroxide at 60 °C. BDMP reacted with PDMS in a novel controlled copolymerization reaction within a good yield. PPO itself composed of aromatic ether resembles a number of vital biomaterials and vitamins [1, 4, 5]. A variety of synthetic polymers has been reported from the random reactions at ortho and para positions of the aromatic nuclei in the basic conditions. None of these reactions provide the necessary controlled reactions required in the new areas of polymer synthesis. The coupling reactions of BDMP with displacement of bromine reported earlier does not specify the role of the solid surface in potassium ferricyanide, lead oxide, or cuprous chloride; in addition, the competitive dimer formation 3,3'-5,5' tetramethyl 4-4' diphenoquinone from the reaction in para position of aromatic nuclei poses a serious problem for the bulky alkyl groups in the aromatic nuclei in the polymerization or copolymerization reactions. In addition, low reactivity of dihalo or trihalo dialkyl monomers have limited the oxidative coupling reactions of the halo dialkyl phenols [1–3]. Recently, reactivity of monomers has been important in a phase transfer radical polymerization, where reactivity of the monomer has been assisted by other alkyl phenols like trimethyl phenol [4–7]. Displacement of halogen atoms in BDMP has been extensively used by us and others to make a flame-retardant polyether with high glass transition temperature [1, 2]. Cross linkings, dimer formations, and low reactivity of BDMP are reported to yield polymers and copolymers with very low molecular weights, limiting its use and applications. This led us to explore a new mechanism to increase reactivity of this monomer in the polymerization reaction and synthesis of new polymers with new properties. This chapter describes the controlled radical reaction of BDMP in a cobalt-mediated system. As will be explained, BDMP monomer is converted to an active radical with benzoyl peroxide initiator in the presence of Co(acac)$_2$/DMF catalyst in the copolymerization reactions [3]. The presence of bromine in the aromatic ring was used to provide cobalt-mediated system to copolymerize this monomer with PDMS. Presence of the radical in this system increased reactivity of the monomer and allowed us to synthesis new copolymers with a number of monomers with new properties [4–7]. Some of the limitations that had restricted the direct application of PPO as such is its high glass transition temperature (T_g = 220 °C), poor melt flow and the susceptibility of its methyl group to thermal oxidation. PPO crystallizes only with difficulty and behaves as an amorphous polymer during melt processing. To overcome the processing difficulty, usually PPO is copolymerized with other polymers like polystyrene, polycarbonates and PDMS for commercial applications [4]. Polycondensation of BDMP in a two-phase system with an oxidizing agent, takes place through a single electron mechanism, has been demonstrated by Price and co-workers [8–11]. Recently, we have become interested in cobalt-mediated controlled radical polymerization to prepare precursors for the synthesis of PPO based copolymers with St, VAc, MMA initiated with Br-PDMS-Br as macroinitiator [Silicon]. Polymerization of BDMP is strongly catalyzed by Co (acac)$_2$/DMF as a catalyst/ligand complex, making the cobalt-mediated system suitable for conversion of this monomer to a high polymer. The control mechanism was extended to homopolymerization of this monomer in Scheme 4.1.

Controlled radical copolymerization of PDMS and 4-bromo-2,6-dimethyl phenol in the presence of Co(acac)$_2$ catalyst with DMF ligand is shown Scheme 4.2.

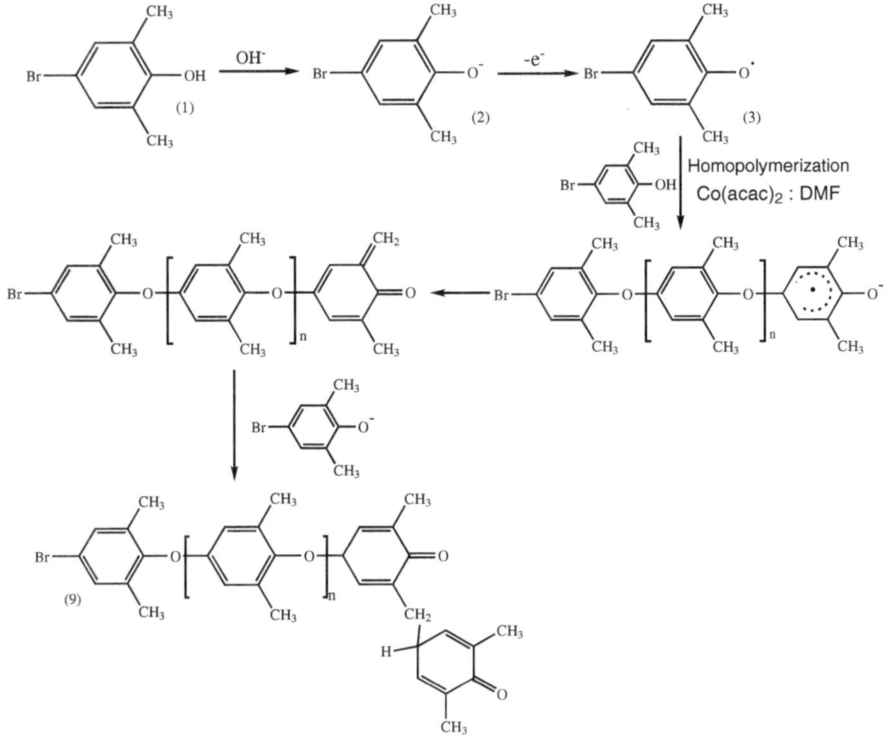

Scheme 4.1 Controlled radical polymerization of 4-bromo-2,6-dimethyl phenol in the presence of Co(acac)$_2$ catalyst with DMF ligand

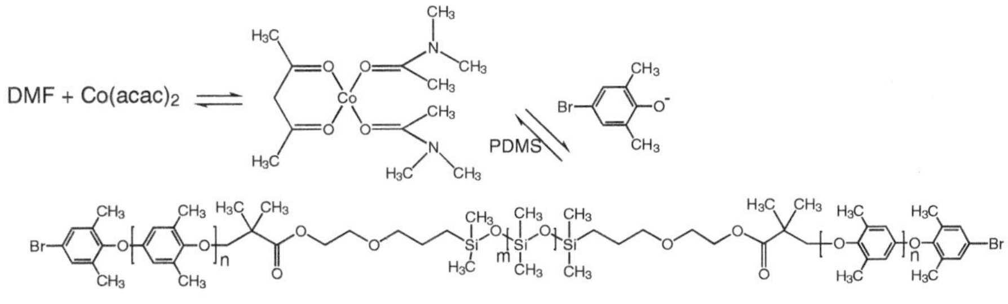

Scheme 4.2 Controlled radical copolymerization of poly(dimethyl siloxane) and 4-bromo-2,6-dimethyl phenol in the presence of Co(acac)$_2$/DMF as catalyst/ligand complex

4.2 PDMS Macroinitiator (Br-PDMS-Br)

^{1}H-NMR of the functionalized PDMS (Scheme 4.3) was used to verify the quantitative modification of the end groups to the terminal bromine atoms (Fig. 4.1). GPC analysis showed M_n = 14,800, with a MWD = 1.93, maximum yield determined gravimetrically is 82 %. ^{1}H-NMR (CDCl$_3$) indicated δ = 0.0–0.3 ppm for protons of methyl groups of -Si(CH$_3$)$_2$O, δ = 2.0 ppm for methylene group next to the bromide [6, 10, 11].

4.3 BDMP Homopolymer and Copolymer

The effect of the initiator, catalyst, KtButOH with Co(acac)$_2$: DMF catalyst has been studied in a number of reactions. In the absence of initiator, the polymer did not form. The reactions yield increased by adding the initiator content (BPO = 2.95 mol L^{-1}, yield = 67 %; and BPO = 3.95 mol L^{-1}, yield = 82 %). The critical role of Co(acac)$_2$: DMF catalyst indicated that in addition to the BPO initiator, the DMF ligand was needed to activate Co(acac)$_2$ catalyst to achieve a high yield of polymer with a very low molecular weight distribution. The same effect was observed with the copolymerization of 4-bromo-2,6-dimethylphenol with poly(dimethylsiloxane), and the presence of the initiator was essential for the reaction. In this case, the copolymerization yield also depends on the initiator (BPO = 2.95 mol L^{-1}, yield = 64 %; and BPO = 3.95 mol L^{-1}, yield = 85 %).

4.3.1 Characterization of the BDMP Homopolymer and Copolymer

The ^{1}H-NMR microstructure of BDMP in the presence of BPO initiator and Co (acac)$_2$/DMF catalyst is shown in Fig. 4.2. The microstructure of the polymer at

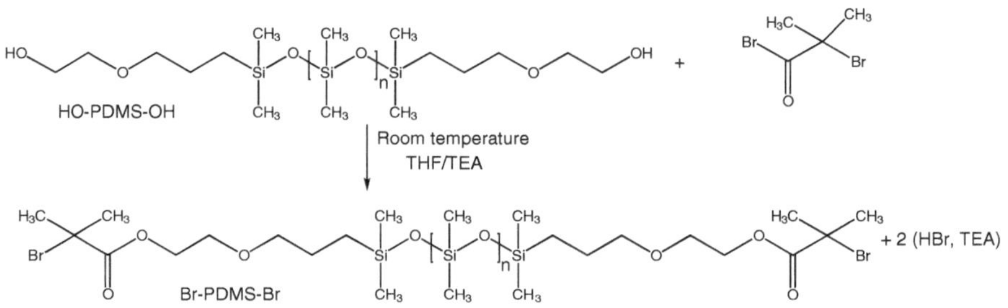

Scheme 4.3 Reaction scheme for the synthesis of bis(2-bromoisobutyrate)-terminated PDMS macroinitiator from bis(hydroxyalkyl)-terminated PDMS

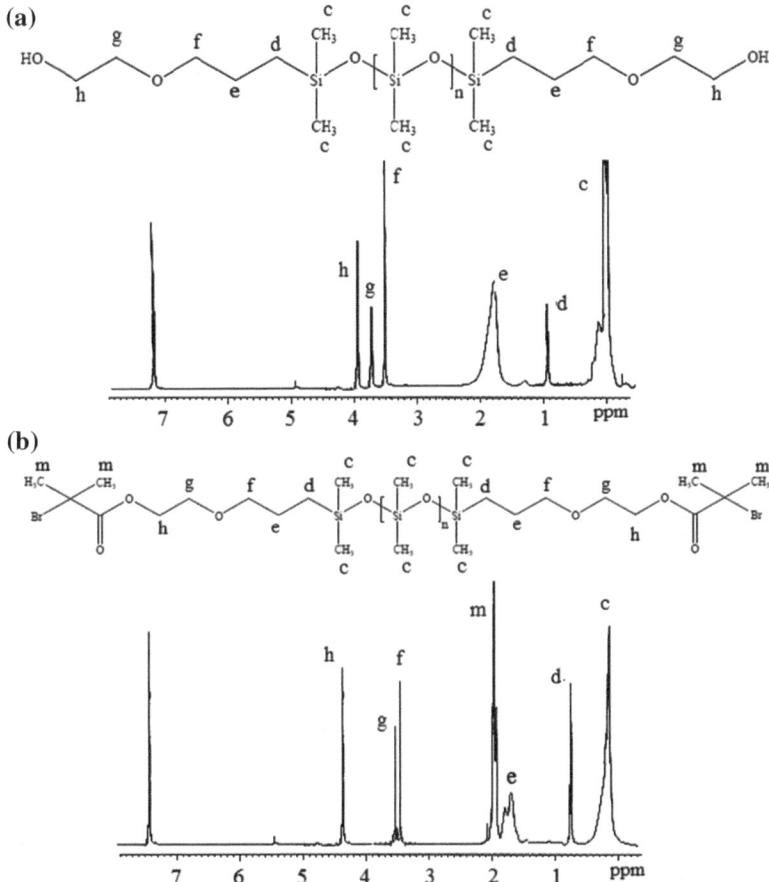

Fig. 4.1 ^{1}H-NMR spectra of HO-PDMS-OH (**a**) and Br-PDMS-Br macroinitiator (**b**) in the CDCl$_3$ solvent

Fig. 4.2 ^{1}H-NMR of PPO homopolymer at high yield

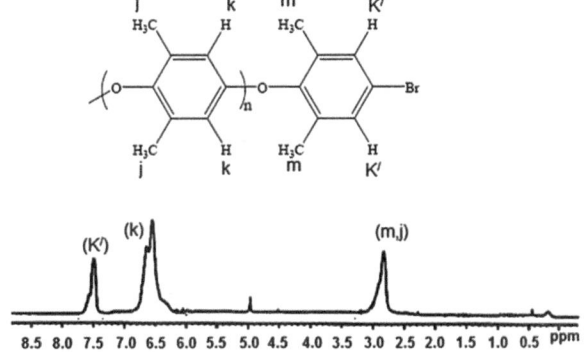

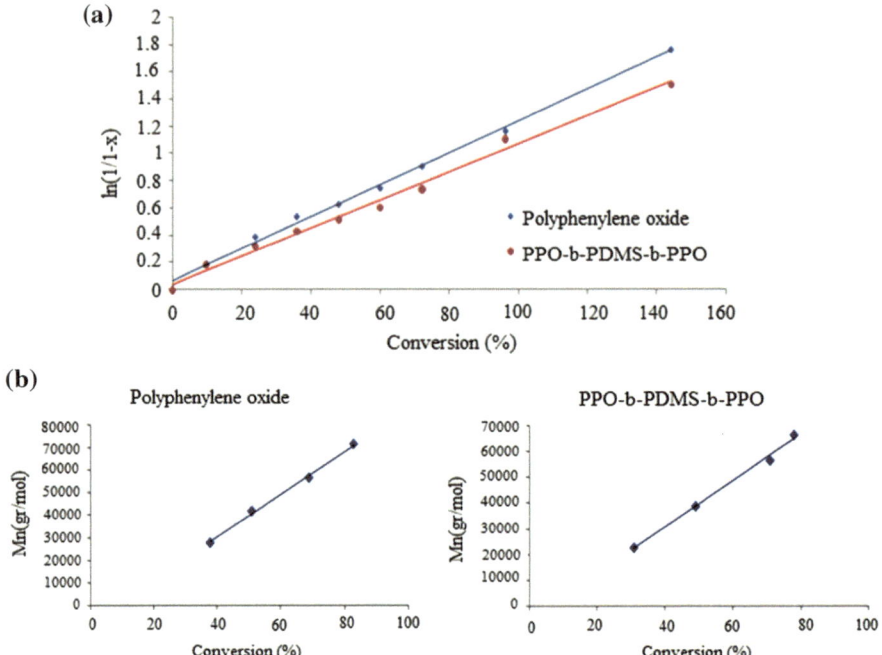

Fig. 4.3 **a** Time dependence of $\ln[M]_0/[M]$ (*M* monomer) **b** dependence of the polymer Mn on the monomer conversion of the PPO homopolymer and PPO-b-PDMS-b-PPO copolymer initiated with BPO initiator at 60 °C

high yield indicated that the catalyst is involved in the controlled radical polymerization (CRP) and formation of the starting unit. Three major peaks related to aromatic protons appeared at 7.5 ppm. The end-group protons and methine proton adjacent to terminal bromine of BDMP and chain unit protons appeared at 6.5 ppm.

Methyl protons of polymer unit were assigned to 2.72 ppm, the peak intensities at 6.47 and 6.34 ppm were related to methyl protons of polymer unit (k) (Fig. 4.2) [4, 6]. Copolymerization of BDMP with poly(dimethyl siloxane), which was initiated by BPO with $Co(acac)_2$: DMF catalyst at 60 °C for 6 h, was studied further to confirm the first-order kinetic of BDMP homo and copolymerization in controlled radical polymerization. The first-order reaction also suggested that the reaction is a controlled radical copolymerization. These results indicated that the controlled radical polymerizations are essentially linear and they are within allowable range of experimental error. The controlled reaction was confirmed further by the average molecular weight of polymers and copolymers. The measured $\bar{M}_n$ indicated that monomer conversion increased linearly with $\bar{M}_n$ (Fig. 4.3).

The linear kinetics of the conversion and the increase in molecular weight of polymer and copolymer are consistent with constant concentration proposed for the controlled radical propagation step of reaction. The high molecular weights of polymers and copolymers with narrow distribution indicated that radical termination reactions were not significant within the time scale of reactions. Since the molecular weight distribution remained low during the polymerization, nearly all the chains were expected to have been started and grow simultaneously during the equilibrium step.

4.3.2 Characterization of PPO-b-PDMS-b-PPO Copolymer

The microstructure of PPO-b-PDMS-b-PPO copolymer confirmed the controlled reaction, indicated and characterized by ^{1}H-NMR in Fig. 4.4. A signal at about 7.09 ppm corresponds to terminal aromatic proton (a) adjacent to terminal bromine of BDMP [4, 6]. Peaks of methylene protons (c) in the neighborhood of BDMP repeating unit appeared at 6.4 ppm. Methyl protons (d, b) of the monomer appeared at 2.14 ppm. Hydrogen atoms of methylene end chains (f) appeared at 3.61 ppm, whereas methyl group of acetate is at 2.11 ppm (k). The methine proton of monomer unit (h) indicated a peak about 5.2 ppm [3, 4]. A signal at 0.0–0.2 ppm associated with CH_3-Si methyl protons of PDMS.

GPC results indicated narrow molecular weight distributions of the polymer and copolymer of BDMP monomer and vinyl acetate. The experimental molecular weights and distribution of the synthesized polymer (PPO) and PPO-b-PDMS-b-PPO copolymers are shown in Table 4.1.

Thermal properties of the PPO homopolymer and PPO-b-PDMS-b-PPO copolymer were analyzed by differential scanning calorimeter (DSC) (Fig. 5.4). The sample was heated to 300 °C at 10 °C/min. The DSC thermograms of PPO (Fig. 4. 5a) shows the first transition at 220 °C, while for the second segment the distinct glass transitions at −120 °C is the characteristic of second block or poly(dimethyl siloxane) in the copolymer (Fig. 4.5b). The first transition at about −80 °C was assigned to the glass transition temperature of PDMS segment ($T_{g,PDMS}$) [3] and the second one at 193 °C to the PPO segment ($T_{g,PPO}$) of the block copolymer [4, 6].

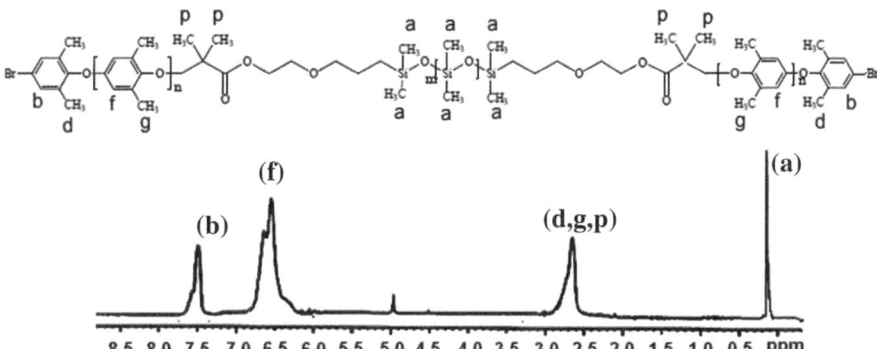

Fig. 4.4 ^{1}H-NMR of PPO-b-PDMS-b-PPO copolymer initiated with BPO with Co(acac)$_2$ at 60 °C for 6 h

Table 4.1 Summary of the results obtained from GPC analyses for PPO homopolymer and PPO-b-PDMS-b-PPO copolymer initiated with BPO with Co(acac)$_2$: DMF catalyst at 60 °C for 6 h

No.	Reaction	Time (h)	$M_{n,GPC}$ (gr/mol)	PDI
A	Polyphenylene oxide	72	41,800	1.31
B	Polyphenylene oxide	96	56,500	1.25
C	Polyphenylene oxide	144	71,300	1.18
D	PPO-b-PDMS-b-PPO	72	49,600	1.35
E	PPO-b-PDMS-b-PPO	96	69,800	1.25
F	PPO-b-PDMS-b-PPO	144	78,400	1.14

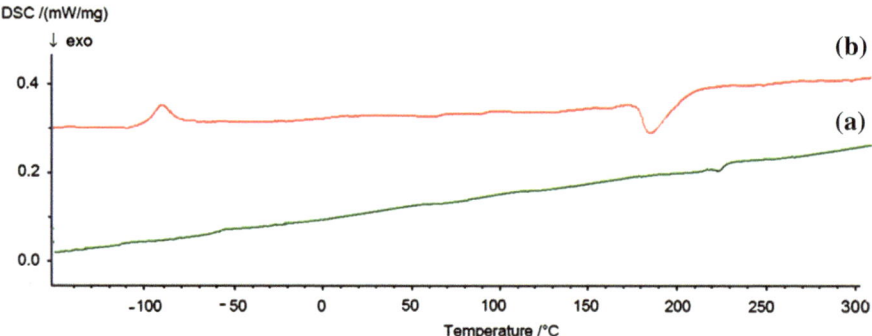

Fig. 4.5 DSC thermograms for the (**a**) PPO homopolymer and (**b**) PPO-b-PDMS-b-PPO copolymer initiated with BPO with Co(acac)$_2$ from −150 to 300 °C with heating rate of 10 °C/min

4.3.3 Reactivity Ratio of PDMS and BDMP

The chemical composition of the copolymers depends on the relative reactivity between the two monomers. Monomer reactivity ratios are important for quantitatively predicting the copolymer composition for any starting feed and for understanding the kinetic and mechanistic aspects of copolymerization. The reactivity ratio was determined for the comonomer pairs PDSM and BDMP.

Kelen-Tüdös, Mayo-Lewis and Fineman-Ross linear methods were used to determine the monomers chemical reactivity of PDMS and BDMP in controlled radical polymerization. The synthesized block copolymers showed a relatively narrow molecular weight distribution specific of the controlled polymerization reactions of CMRPM. A broadening trend of PDI in block copolymers was observed with polymerization temperature. Characterization of molecular structure, number average molecular weight, and molecular weight distribution are also reported for novel block copolymers by ^{1}H-NMR spectroscopy, FTIR spectrometery, and gel permeation chromatography (GPC) techniques.

The values of r_1 and r_2 which were estimated for PDMS and BDMP monomer pairs from Mayo-Lewis ($r_{PDMS} = 0.293$, $r_{BDMP} = 0.485$), Kelen-Todous ($r_{PDMS} = 0.289$, $r_{BDMP} = 0.476$) and Finemann-Rose ($r_{PDMS} = 0.299$, $r_{BDMP} = 0.481$) indicates that the growing radicals tend to addition to BDMP monomer.

4.4 Heptablock Copolymers Based on PDMS

4.4.1 Synthesis of New Block Copolymers via Controlled Radical Polymerization

The first-order kinetic plot of controlled radical polymerization indicated constant concentration of propagating radical species and radical termination reactions are not significant on the time scale of this reaction. This is reported further with linear increase of molecular weights with conversion, indicating that the number of chains is constant during the reaction. The controlled polymerization indicated a narrow molecular weight distribution with good agreement between theoretical and experimental molecular weights [12].

4.4.2 Characterization of New Block Copolymers via Controlled Radical Polymerization

^{1}H-NMR spectra of the block copolymers have been shown in Fig. 4.6. All signals appeared in the ^{1}H-NMR spectra were assigned to the corresponding monomers. It is clear from Fig. 4.6a that block copolymers have successfully been synthesized. The characteristic PDMS signal of CH_3-Si methyl protons at about 0.0–0.2 ppm was still present in both spectra. The signal seen at about 7.09 ppm corresponds to the methine proton adjacent to terminal bromine of BDMP [12]. Peaks of methylene protons of in neighborhood of BDMP (k) appear at 5.4 ppm. The methyl protons of PPO (r) appear 2.9 ppm [12].

Hydrogen atoms of methylene at the ending of both chains (d) are attributed to the peak of 3.61 ppm, whereas methyl group of the acetate is revealed at 2.11 ppm (y). Protons of the methylene for repeating units (f), end of chains (g) give broad signals at 1.5 to about 2.0 ppm through ^{1}H-NMR spectra. The methine proton of the first VAc unit (s) has a peak about 5.2 ppm [12, 13]. Additional broad signals at about 6.6–7.1 and 1.2–2.3 ppm were observed in Fig. 4.6b, which was assigned to the phenyl ring protons and aliphatic protons, respectively of the PSt segment. Additional signals were also observed in the ^{1}H-NMR spectra of two other triblock copolymers (Fig. 4.6b) which were assigned to the corresponding protons of monomers (MMA) incorporated into the triblock copolymer chain. The signals at 0.7–1.25 ppm are assigned to the methyl protons (–CH_3) of MMA which is shown in Fig. 4.6b [12, 13].

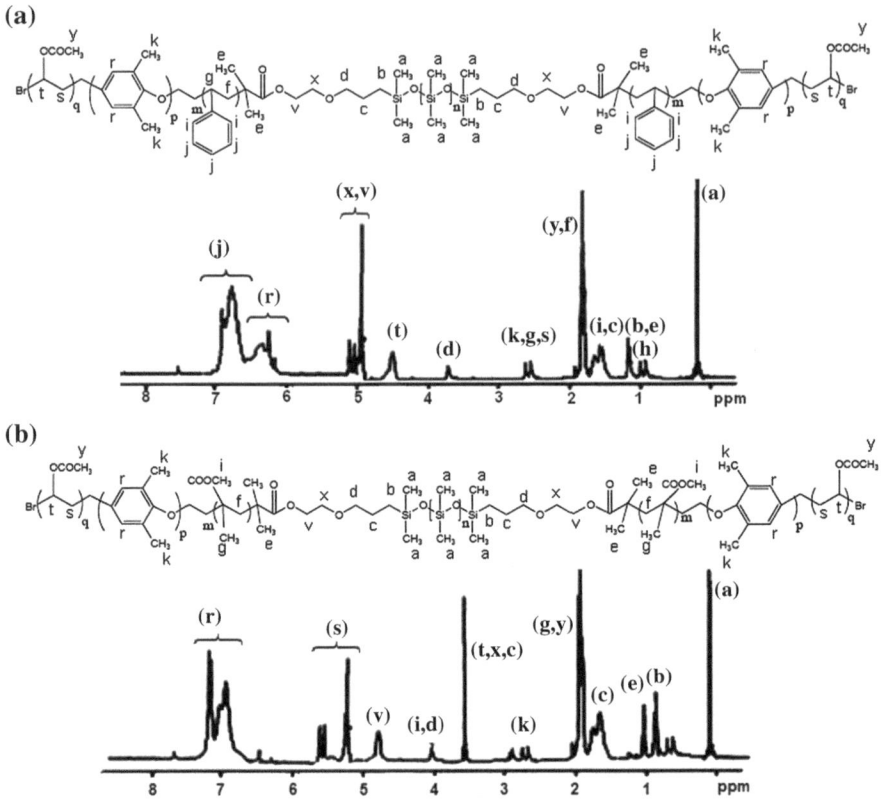

Fig. 4.6 ¹H-NMR of **a** PVAc-b-PPO-b-PMMA-PDMS-b-PMMA-b-PPO-b-PVAc and **b** PVAc-b-PPO-b-PSt-PDMS-b-PSt-b-PPO-b-PVAc initiated with BPO at 60 °C

The FTIR spectrum of the heptablock copolymers are provided in Fig. 4.7. It can be observed that the band at 1435.3 cm^{-1} corresponds to an asymmetrical bending vibration (CH_3) of methyl group of the PMMA. Besides, the vibrational bands observed at 2,855 and 1370.7 cm^{-1} are ascribed to CH_3 symmetric stretching and symmetric bending vibrations of pure PVAc, respectively. In addition, a strong band at 1718.3 and 1726.9 cm^{-1} can be attributed to the carbonyl group of PMMA and PVAc segments, respectively. Also, the frequency shift of the peak due to C-O band of PMMA around 1149.9 cm^{-1} for the sample of the blend implies that there is a specific interaction between PMMA and PVAc. The band at 1,242 cm^{-1} is assigned to -O-$COCH_3$ bond of PVAc segment [12–14].

The peak at 2,951 cm^{-1} is corresponded to the asymmetric stretch vibration of -C-H (CH_3) bond of both PVAc and PMMA segments in block copolymer. The characteristic spectrum of PDMS can be observed at Si-O-Si in 1,100–1,000 cm^{-1}, CH_3 in 1,261 cm^{-1}, C-Si-C in 800 cm^{-1}. The characteristic PDMS absorption bands of "Si-O-Si" at about 1,020 cm^{-1}, "CH_3" at about 1,260 cm^{-1} and "C-Si-C"

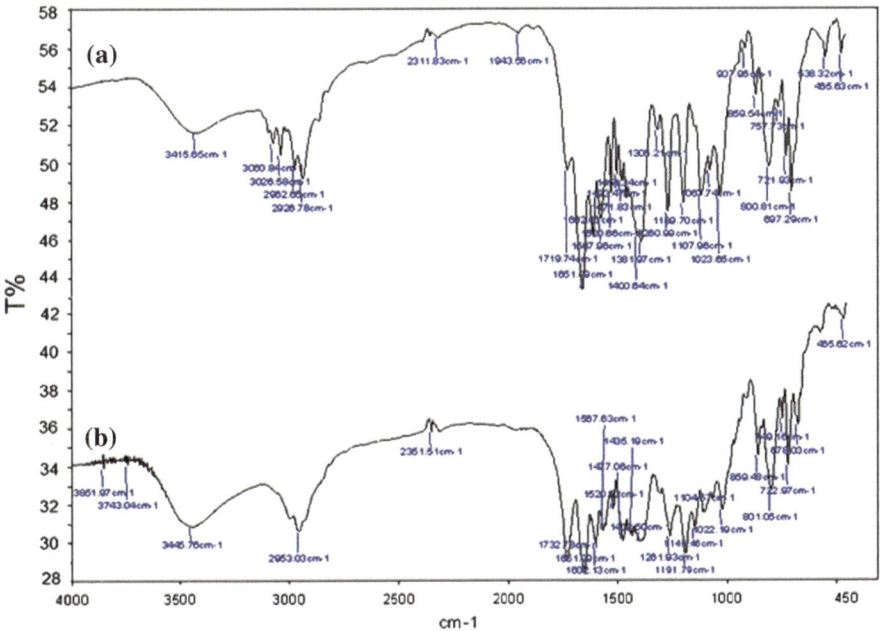

Fig. 4.7 FTIR of **a** PVAc-b-PPO-b-PMMA-b-PDMS-b-PMMA-b-PPO-b-PVAc and **b** PVAc-b-PPO-b-PSt-b-PDMS-b-PSt-b–PPO-b-PVAc initiated with BPO at 60 °C

Table 4.2 Summary of the results obtained from GPC analyses for the PPO based copolymer initiated with BPO with Co(acac)$_2$ at 60 °C for 6 h

No.	Reaction	M_n	PDI
A	PVAc-b-PPO-b-PSt-b-PDMS-b-PSt-b-PPO-b-PVAc	79,000	1.1
B	PVAc-b-PPO-b-PMMA-b-PDMS-b-PMMA-b-PPO-b-PVAc	77,800	1.13

at about 800 cm^{-1} are still present in the Fig. 4.7. Moreover, the characteristic PSt absorption bands appeared at 696, 758, 2,924, 3,028, and 3,062 cm^{-1} (Fig. 4.7b) and the characteristic PMMA absorption bands appeared at 1,740 and 1,178 cm^{-1} (ester group characteristic), indicating the successfully synthesis of PDMS-based block copolymers [12, 13].

GPC traces for the progress of the PPO block copolymers are shown in Table 4.2. It can be implied from the GPC results that Co(acac)$_2$ has controlled the reaction [12].

The DSC thermogram spectrum of synthesized PVAC-b-PPO-b-PVAc, PVAc-b-PPO-b-PSt-PDMS-b-PSt-b-PPO-b-PVAc and PVAc-b-PPO-b-PMMA-PDMS-b-PMMA-b-PPO-b-PVAc block copolymer is also shown in Fig. 4.8. Figure 4.8a showed two distinct glass transitions at 38 °C for PVAc and at 190 °C for PPO (Fig. 4.8a) while 4 distinct glass transitions are observed on the curve of block copolymer (Fig. 4.8b), the transition at about 38 °C is due to the glass transition

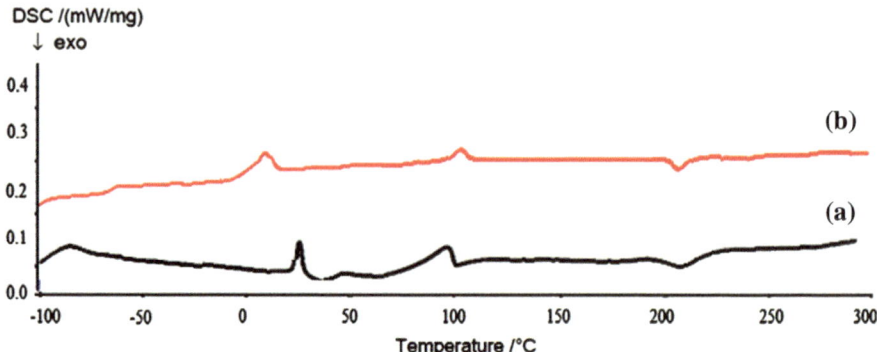

Fig. 4.8 DSC thermogram for the **a** PPO-b-PVAc-b-PSt-PDMS-b-PSt-b-PVAc-b-PPO and **b** PPO-b-PVAc-b-PMMA-PDMS-b-PMMA-b-PVAc-b-PPO initiated with BPO with Co(acac)$_2$ from −100 to 300 °C with heating rate of 10 °C/min

temperature of PVAc segment ($T_{g,PVAc}$) [7, 8, 12, 13], the transition at about −80 °C is due to the glass transition temperature of PDMS segment ($T_{g,PDMS}$), the transition at about 88 °C (Fig. 4.8b) is due to the glass transition temperature of PSt segment ($T_{g,PSt}$), and the other at about 108 °C (Fig. 4.8) corresponds to the PMMA segment ($T_{g,PMMA}$) [12, 13]. The glass transition temperatures of the PVAc and PDMS segments in both block copolymers were similar.

4.5 Summary

A new class of PPO based block copolymers were synthesized through cobalt mediated radical polymerization in the presence of Co(acac)$_2$: DMF as catalyst: ligand complex. The Co(acac)$_2$ catalyst can provide a new syndiotactic route to produce block copolymers having controlled chain structures. The composition analysis by ^{1}H-NMR, FTIR, DSC and GPC confirmed a well-defined triblock and pentablock copolymers microstructures. The narrow dispersity indices of block copolymers implying the effect of living/controlled characteristic of Co(acac)$_2$: DMF and its effect in synthesis of new PPO based block copolymers with high thermal stability.

The novel synthesis of PPO copolymers with various monomers introduced a controlled radical polymerization process that allows the use of BDMP as an aromatic ether nucleus in a number of new vinyl monomers synthesis like vinyl acetate, styrene and methyl methacrylate. The monomodal distribution and very narrow molecular weight distribution with high stability further confirmed the controlled nature of the radical copolymerization of BDMP. The new block copolymers are considered to widen the range of stable aromatic ether block copolymers with specific microstructures and molecular weights with a narrow distribution for the current design of membranes and sensors.

References

1. Semsarzadeh MA, Price CC (1977) Macromolecules 10:2–11
2. Price CC, Semsarzadeh MA, Nguyen TBL (1975) J Appl Polym Sym 26:319–326
3. Semsarzadeh MA, Amiri S (2012) J Chem Sci 24:521–527
4. Percec V, Wang JH (1991) J Polym Sci A Polym Chem 29:63–82
5. Feng SS, Chien S (2003) Chem Eng Sci 58:4087–4114
6. Percec V, Wang JH (1991) Polym Bull 25:33–40
7. Hunter WH, Whitney RB (1932) J Am Chem Soc 54:1167–1173
8. Staffin GD, Price CC (1960) J Am Chem Soc 82:3632–3634
9. Price CC, Chu NS (1962) J Polym Sci 61:135–141
10. Price CC (1974) Acc Chem Res 7:294–301
11. Percec V, Shaffer TDJ (1986) Polym Sci Part C Polym Lett 24:439–446
12. Semsarzadeh MA, Amiri S (2013) Silicon. doi:10.1007/s12633-013-9161-3
13. Semsarzadeh MA, Amiri S (2013) J Inorg Organomet Polym 23:432–438
14. Semsarzadeh MA, Abdollahi M (2012) J Appl Polym Sci 4:2423–2430

Chapter 5
Novel Thermoreversible Block Copolymers: Silicone Macroinitiator in Atom Transfer Radical Polymerization

Abstract Organic–inorganic pentablock copolymers have been synthesized via atom transfer radical polymerization (ATRP) of styrene (St) and vinyl acetate (VAc) monomers at 60 °C using CuCl/PMDETA as a catalyst system initiated from bis(boromoalkyl)-terminated poly(dimethylsiloxane) (PDMS)/γ-cyclodextrin macroinitiator (Br-PDMS/γ-CD). Br-PDMS-Br was reacted with γ-CD in different conditions with inclusion complexes being characterized through hydrogen nuclear magnetic resonance (^{1}H-NMR) and differential scanning calorimetry (DSC). Resulting Br-PDMS-Br/γ-CD inclusion complexes were taken as macroinitiators for ATRP of St and VAc. Well-defined poly(styrene)-b-poly(vinyl acetate)-b-poly (dimethylsiloxane/γ-cyclodextrin)-b-poly(vinyl acetate)-b-poly(styrene) (PSt-b-PVAc-b- PDMS/c-CD-b-PVAc-b-PSt) pentablock copolymer was characterized by ^{1}H-NMR, gel permeation chromatograph (GPC) and DSC. PSt-b-PVAc-b-PDMS/ c-CD-b-PVAc-b-PSt pentablock copolymer can undergo a temperature-induced reversible transition upon heating of the copolymer complex from white complex at 22 °C to green complex in 55 °C which characterized with XRD and ^{1}H-NMR.

Keywords γ-Cyclodextrin (γ-CD) · Poly(dimethylsiloxane) (PDMS) · Poly(vinyl acetate) (PVAc) · Atom transfer radical polymerization (ATRP) · Sonic energy · Inclusion complex · Thermosensitive block copolymer

5.1 Introduction

The majority of silicone based polymers is poly(dimethyl siloxanes). Copolymers containing poly(dimethylsiloxane) (PDMS) have received considerable attention due to their unique properties, such as very low glass transition temperature, low surface energy, low solubility parameter, and physiological inertness. Some of their specialty applications are in the fields of biomaterials, surfactants and new thermoreversible polyrotaxanes. Due to their unique structures and properties, polyrotaxanes represent an important addition to the repertoire of polymer architectures. Recently, the supramolecular assemblies of rotaxane type received a renewed interest, analogous to the one on noncovalent interactions in biological systems.

Polyrotaxanes have attracted much attention in the past decades for their great potential as stimulus-responsive materials which can produce PDMS based thermoreversible block copolymers. They are usually constructed by threading molecular rings (or host molecules) onto a macromolecular chain via host-guest interaction, followed by capping the chain ends with bulky groups. Among all the host molecules, cyclodextrins (CDs) are frequently applied in polyrotaxane system mainly due to their bioavailability and low cytotoxicity. This chapter describes the atom transfer radical polymerization (ATRP) and cobalt mediated radical polymerization (CMRP) synthesis of novel thermoreversible block copolymers based on PDMS macroinitiators with various monomers. Polyrotaxane based PDMS have been synthesized via 2 methods. In the first method, Polyrotaxane have been synthesized via ATRP of styrene (St), methyl methacrylate (MMA) and vinyl acetate (VAc) monomers at 60 °C using CuCl/N,N,N',N'',N''-pentamethyldiethylenetriamine (PMDETA) as a catalyst system initiated from bromoalkyl-terminated PDMS/cyclodextrins macroinitiator (Br-PDMS-Br/γ-CD). Inclusion complex of Br-PDMS-Br and with γ-CD was characterized through ^{1}H-NMR, GPC and DSC. Then Br-PDMS-Br/γ-CD was used as macroinitiators for ATRP of St and VAc. Well-defined poly(styrene)-b-poly(vinyl acetate)-b-poly(dimethyl siloxane/γ-cyclodextrin)-b-poly(vinyl acetate)-b-poly(styrene) (PSt-b-PVAc-b-PDMS/γ-CD-b-PVAc-b-PSt) pentablock copolymer was characterized by ^{1}H-NMR, GPC and DSC. There was a very good agreement between the number-average molecular weight calculated from ^{1}H-NMR spectra and that of theoretically calculated. Novel supramolecular block copolymers can undergo a temperature-induced reversible transition upon heating-cooling of the copolymer complex from white complex at 22 °C to green complex in 55 °C which characterized with XRD and ^{1}H-NMR. XRD showed a change in crystalinity percent of St peak with changing the temperature. PDMS based polyrotaxane are characterized using FTIR, ^{1}H-NMR, DSC, and GPC. There was a good agreement between the number-average molecular weight calculated from ^{1}H-NMR spectra and that of theoretically calculated which indicated controlled manner of these systems.

PDMS macroinitiator could be used in two forms:

1. Used inclusion complex between γ-CD and Br-PDMS-Br as macroinitaitor in ATRP of various monomers
2. Synthesis of pentablock copolymers via ATRP and then reacted with γ-CD in a CMRP

In an effort to produce materials composed of silicones with more desirable mechanical properties, block, graft, and network copolymers containing PDMS segments have been reacted with γ-cyclodextrin. When inclusion complexes of CDs with silicon-containing polymers are formed, they are new organic-inorganic hybrids with exact stoichiometric relationships which showed stimuli-responsive properties with temperature and pH. Polyrotaxane based on block copolymers can undergo a temperature-induced reversible transition upon heating of the copolymer complex from a white complex at 22 °C to a green complex in 55 °C which characterized with XRD and ^{1}H-NMR. XRD showed a change in crystalinity percent of St peak with changing the temperature. These pentablock block copolymers revealed thermosensitive micelle formation behavior in the solid state. As temperature increased, the CDs were dehydrated while the color of powder was changed from white to green through hydrophobic-hydrophobic interactions. Further increase of the temperature caused the phase transition related to PDMS to occur on the corona of noncovalently connected micelles. Therefore, the micelles were destabilized which resulted in micelle aggregation and precipitation. The supramolecular system presented here is rather different from the reported systems. This work has formed dual thermoresponsive pentablock copolymers in order to propose a novel supramolecular approach in designing and constructing the non-covalently connected polymeric micelles with modifiable properties.

5.2 Synthesis of Thermoreversible Block Copolymers via First Method

5.2.1 Inclusion Complex of Br-PDMS-Br and γ-CD as Macroinitiator

First-order kinetic polymerization plots of St and PVAc initiated by Br-PDMS-Br/ γ-CD macroinitiator are shown in Fig. 5.1.

The linear fit indicated that the concentration of propagating radical species is constant and radical termination reactions are not significant over time scale of the reaction. It was found that the molecular weights increase almost linearly with conversion, indicating that the number of chains was constant and the chain transfer reactions were rather negligible [1–3]. Another important feature observed for controlled radical polymerization was that the molecular weight distribution decreases with the progress of polymerization, indicating that nearly all the chains start to grow simultaneously [4–8]. Thus from Fig. 5.1, one may conclude that the polymerization was controlled and had narrow molecular weight distribution with a good agreement between theoretical and experimental results of molecular weight (Table 5.1). GPC is an effective analytical technique to discover the structure of polyrotaxanes. If polyrotaxanes are not efficiently end-capped, dethreading of CDs will produce multiple GPC peaks. The GPC curves of polyrotaxanes-based pentablock copolymers (PSt-b-PVAc-b-PDMS/γ-CD-b-PVAc-b-PSt) are depicted in Table 5.1. It is implied that the number of entrapped γ-CD can be modified in this ATRP process [1, 2].

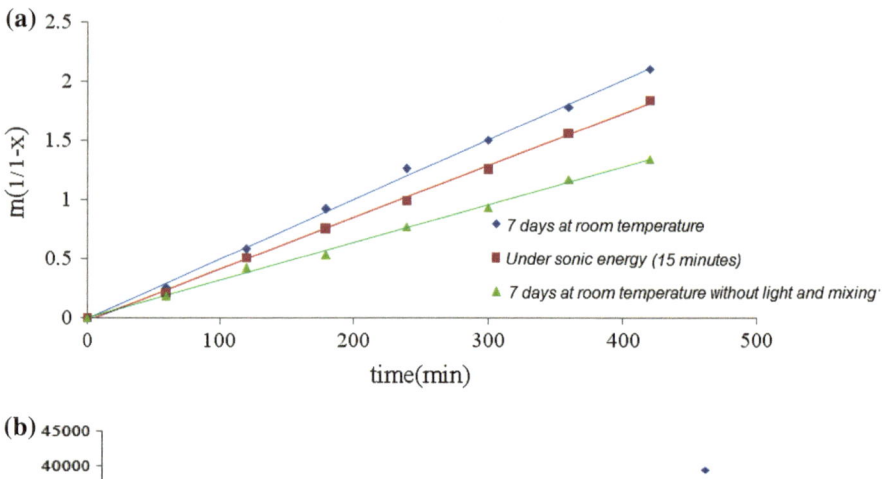

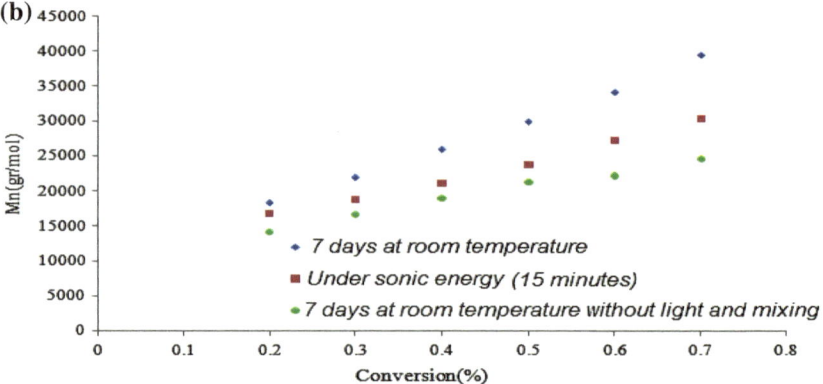

Fig. 5.1 a Time dependence of $\ln[M]_0/[M]$ (*M* monomer); **b** dependence of the pentablock copolymer M_n on the conversion for the ATRP of St and PVAc with Br-PDMS-Br/γ-CD at 60 °C

Table 5.1 GPC and ^{1}H-NMR results of PSt-b-PVAc-b-[γ-CD/PDMS]-b-PVAc-b-PSt initiated with γ-CD/PDMS synthesized from various conditions

Reaction	Reaction	X (%)[a]	$\bar{M}_{n,\,\text{H-NMR}}^{1}$ (g mol^{-1})	$\bar{M}_{n,\text{GPC}}$ (g mol^{-1})	PDI
A	7 days at room temperature without light and mixing	40	41,900	44,100	1.53
B	Under sonic energy (15 min)	56	46,300	45,400	1.37
C	7 days at room temperature	72	48,500	49,500	1.24

[a] Final conversion measured by gravimetric method

5.2.2 Characterization of Thermoreversible Block Copolymers Synthesized via First Method

The thermo-responsiveness of the pentablock copolymers was investigated with
^{1}H-NMR in CDCl$_3$. Figure 5.2 shows the ^{1}H-NMR spectrum obtained for the PSt-
b-PVAc-b-PDMS-b-PVAc-b-PSt at 22 °C (Fig. 5.2a), 55 °C (Fig. 5.2b) and 22 °C
(Fig. 5.2c) [9]. The integral of the peaks associated with protons adjacent to the
styrene and vinyl acetate (6.5–7 ppm and 1.2–2.1 ppm respectively) were observed
to decrease with increasing temperature from 22 to 55 °C relative to the integral of
the protons in the PMDS (Fig. 5.2b). This reflects the mobility of the γ-cyclodextrin
in block copolymer structure, upon decreasing temperature to 22 °C, integral of the
peaks associated with protons adjacent to the styrene and vinyl acetate the
copolymers undergo increase to reach initial integral(Fig. 5.2c).

Figure 5.3 shows the XRD spectra obtained for the PSt-b-PVAc-b-PDMS/γ-CD-
b-PVAc-b-PSt at 22 °C (Fig. 5.3a), temperature increased to 55 °C (Fig. 5.3b) and
temperature decreased from 55 to 22 °C (Fig. 5.3c).

Characteristic peaks of PDMS occurring at 2θ = 8, 13, 18, characteristic peaks of
PVAc occurring at 2θ = 13 and characteristic peaks of PSt occurring at 2θ = 14, 12,
11, 8, 6–17, 25–9. At 22 °C, crystalinity percent of PSt and PVAc 29 and 13 %
respectively (Fig. 5.3a). By increasing temperature from 22 to 55 °C, crystalinity
percent of PSt and PVAc becomes 43 and 9 % respectively (Fig. 5.3b), which
indicated migration of CD from PVAc to PSt. By decreasing temperature from 55
to 22 °C, crystalinity percent of PSt and PVAc becomes 10 and 35 % respectively
(Fig. 5.3c), which indicated migration of CD from PVAc to PMMA [10–15].

5.3 Thermoreversible Block Copolymers Synthesized via Direct Reaction with γ-CD (Second Method)

5.3.1 Pentablock Copolymers via ATRP and Reacted with γ-Cyclodextrin

Synthesized PSt-b-PMMA-b-PVAc-b-PMMA-b-PSt and PSt-b-PMA-b-PVAc-b-
PMA-b-PSt pentablock copolymers were reacted with γ-cyclodextrin in cobalt
mediated radical polymerization [16]. First-order kinetic polymerization plots of
hepta copolymers of PSt-b-PMMA-b-PVAc-b-PDMS-b-PVAc-b-PMMA-b-PSt
and PSt-b-PMA-b-PVAc-b-PDMS-b-PVAc-b-PMA-b-PSt polyrotaxanes are
shown in Fig. 5.4.

The linear fit indicated that the concentration of propagating radical species is
constant and radical termination reactions are not significant over time scale of the
reaction. It was found that the molecular weights increase almost linearly with
conversion, indicating that the number of chains was constant and the chain transfer
reactions were rather negligible [1, 2, 9].

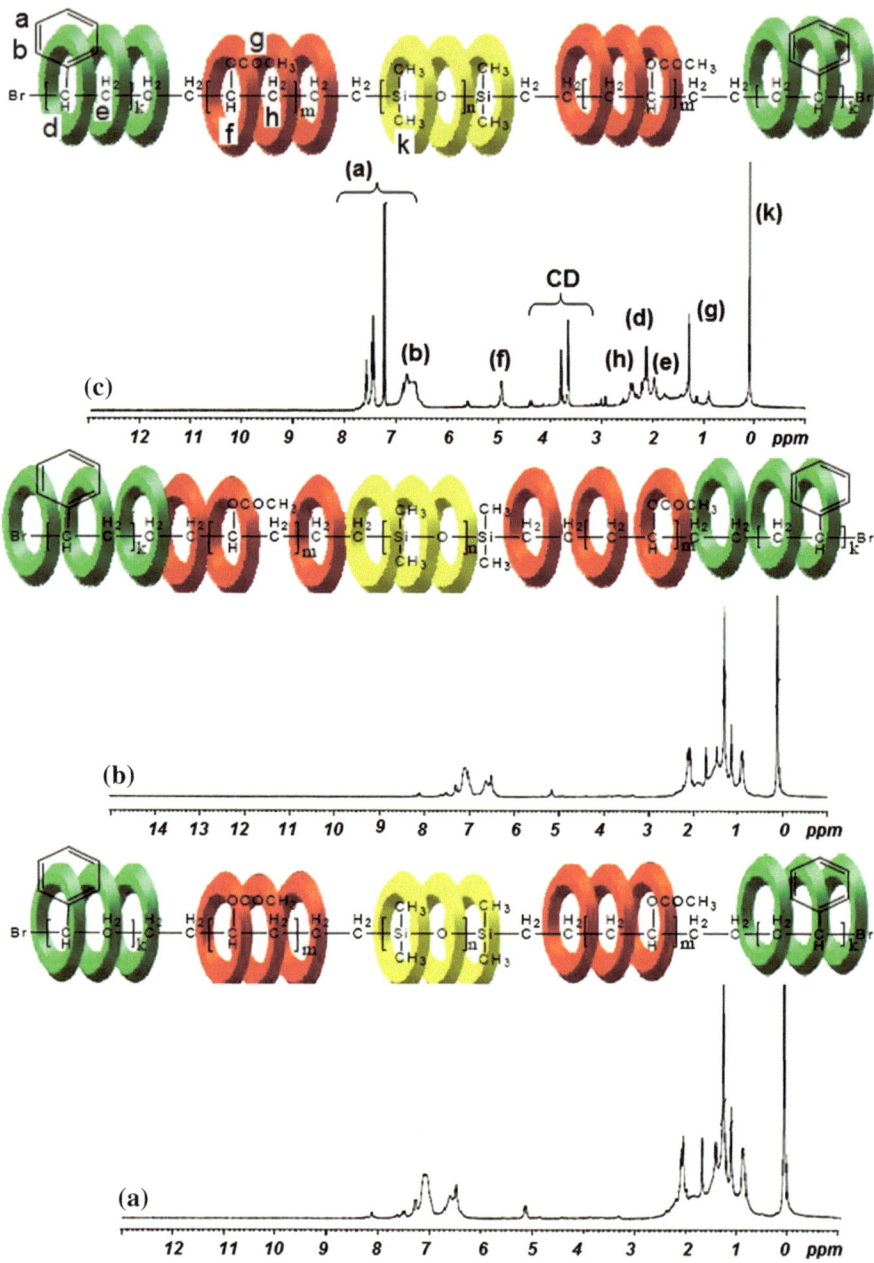

Fig. 5.2 ^{1}H-NMR of thermoreversible PSt-b-PVAc-b-PDMS/γ-CD-b-PVAc-b-PSt block copolymers initiated with inclusion complex of γ-CD/Br-PDMS-Br synthesized (7 days at room temperature) at **a** 22 °C, **b** 55 °C and **c** 22 °C

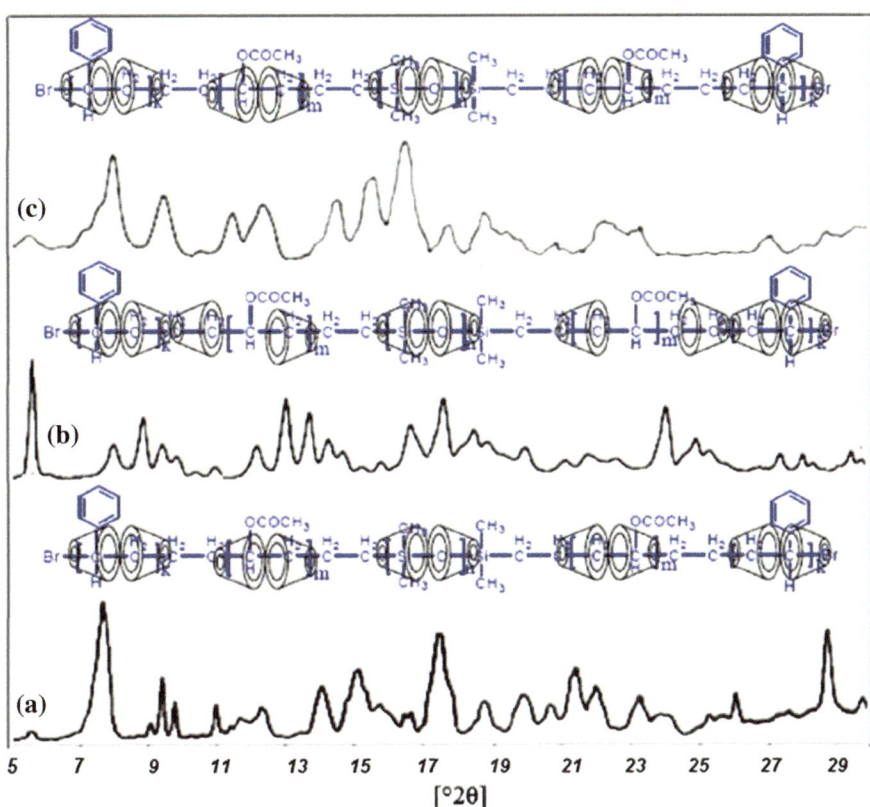

Fig. 5.3 XRD of thermoreversible polyrotaxane of PSt-b-PVAc-b-PDMS/γ-CD-b-PVAc-b-PSt block copolymers initiated with inclusion complex of γ-CD/Br-PDMS-Br synthesized (7 days at room temperature) at **a** 22 °C, **b** 55 °C and **c** 22 °C

5.3.2 Characterization of Thermoreversible Block Copolymers Synthesized via Direct Reaction with γ-CD (Second Method)

GPC is an effective analytical technique to discover the structure of polyrotaxanes. If polyrotaxanes are not efficiently end-capped, dethreading of CDs will produce multiple GPC peaks. The GPC curves of polyrotaxanes-based block copolymers (PSt-b-PMMA-b-PVAc-b-PDMS-b-PVAc-b-PMMA-b-PSt and PSt-b-PMA-b-PVAc-b-PDMS-b-PVAc-b-PMA-b-PSt polyrotaxane) are depicted in Fig. 5.5 (Table 5.2). A nearly symmetrical and unimodal peak with a relatively low polydispersity index of 1.24–1.53 can be clearly observed in Fig. 5.5.

From Table 5.2 and M_n of PSt-b-PMMA-b-PVAc-b-PDMS-b-PVAc-b-PMMA-b-PSt (38,500) and PSt-b-PMMA-b-PVAc-b-PDMS-b-PVAc-b-PMA-b-PSt (28,900)

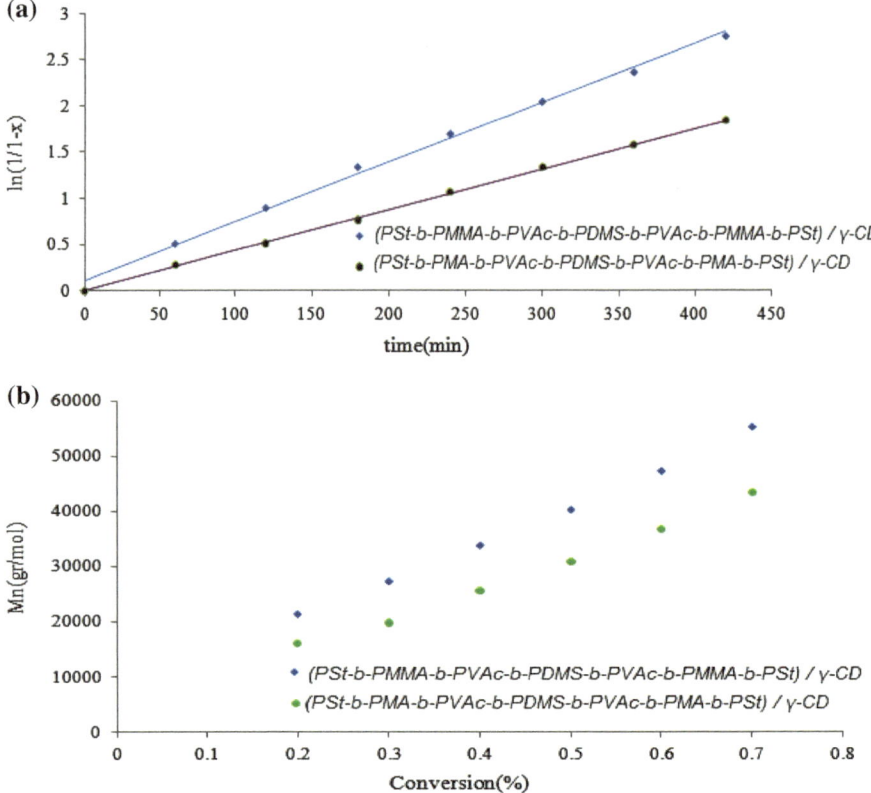

Fig. 5.4 a Time dependence of $\ln[M]_0/[M]$ (*M* monomer); **b** dependence of the block copolymer M_n on the conversion for the ATRP of St, MA, MMA and PVAc with Br-PDMS-Br at 60 °C

and M_n γ-CD (1,270), number of CDs capped in polyrotaxane can be calculated which is shown in Table 5.2 [1, 2].

The thermo-responsiveness of the pentablock copolymers was investigated with ^{1}H-NMR in CDCl$_3$ and XRD at 22 and 55 °C. ^{1}H-NMR and XRD spectra obtained for the PSt-b-PMMA-b-PVAc-b-PDMS-b-PVAc-b-PMMA-b-PSt and PSt-b-PMA-b-PVAc-b-PDMS-b-PVAc-b-PMA-b-PSt based polyrotaxane in thermal cycle from 22 to 55°C (Figs. 5.6, 5.7, 5.8 and 5.9). All signals of the ^{1}H-NMR spectra were assigned to their corresponding monomers and it can be declared that the synthesis of new polyrotaxane based block copolymers have proceeded successfully [1, 8, 17]. The ^{1}H-NMR of polyrotaxane showed a signal at 0.0–0.2 ppm associated with CH$_3$-Si methyl protons of PDMS in addition to some other signals at 2.1–2.3 ppm and 6.6–7.1 ppm corresponding to the OCOCH$_3$ group from PVAc segment and PSt segment, respectively. Signals at 0.8–1.0 ppm, 1.9–2.1 ppm and 3.6 ppm corresponded to CH$_3$, CH$_2$ and COOCH$_3$ group from the PMMA and PMA segment. Signals observed at about 4.5 ppm were attributed to the OH of CDs segment [9, 17].

Fig. 5.5 GPC of a PSt-b-PMMA-b-PVAc-b-PDMS-b-PVAc-b-PMMA-b-PSt and **b** PSt-b-PMA-b-PVAc-b-PDMS-b-PVAc-b-PMA-b-PSt initiated with Br-PDMS-Br at 60 °C in 6 h

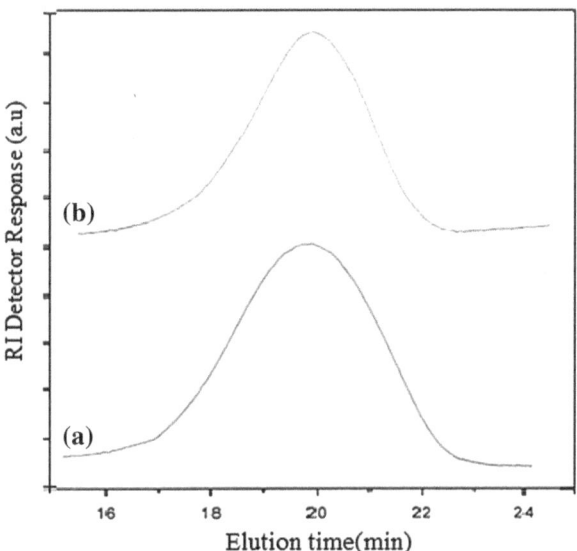

Table 5.2 Results Obtained from GPC and ^{1}H-NMR Analyses for PSt-b-PMMA-b-PVAc-b-PDMS-b-PVAc-b-PMMA-b-PSt and PSt-b-PMA-b-PVAc-b-PDMS-b-PVAc-b-PMA-b-PSt block copolymers

(Co)Polymer	X (%)[a]	$M_{n,GPC}$ (g mol^{-1})	PDI	CD numbers
(PSt-b-PMMA-b-PVAc-b-PDMS-b-PVAc-b-PMMA-b-PSt)/γ-CD	67	55,264	1.21	13
(PSt-b-PMA-b-PVAc-b-PDMS-b-PVAc-b-PMA-b-PSt)/γ-CD	51	43,505	1.34	11

[a] Final conversion measured by gravimetric method

The interaction of cyclodextrin with aromatic end group of the block copolymer, indicated in peaks around 6.0–8.0 ppm shows temperature dependency. The peaks shift during the temperature cycle from 55 to 22 °C indicates temperature induced effect on the complex in a reversible process that dislocates cyclodextrin from the end group of the block copolymer [9]. Figure 5.6 shows the ^{1}H-NMR spectra obtained for the PSt-b-PMMA-b-PVAc-b-PDMS-b-PVAc-b-PMMA-b-PSt at 22 °C (Fig. 5.6a), temperature increased to 55 °C (Fig. 5.6b) and temperature decreased from 55 to 22 °C (Fig. 5.6c). The integral of the peaks associated with protons adjacent to the methyl methacrylate (2.0–2.5 ppm) were observed to decrease which integral of the peaks associated with protons adjacent to the vinyl acetate (1.2–2.1 ppm) were observed to increase with increasing temperature from 22 to 55 °C relative to the integral of the protons in the PMDS (Fig. 5.6b). This reflects the mobility of the γ-cyclodextrin in block copolymer structure, upon decreasing temperature to 22 °C, integral of the peaks associated with protons adjacent to the vinyl acetate the copolymer undergo decrease to reach initial integral (Fig. 5.6c) [18–22].

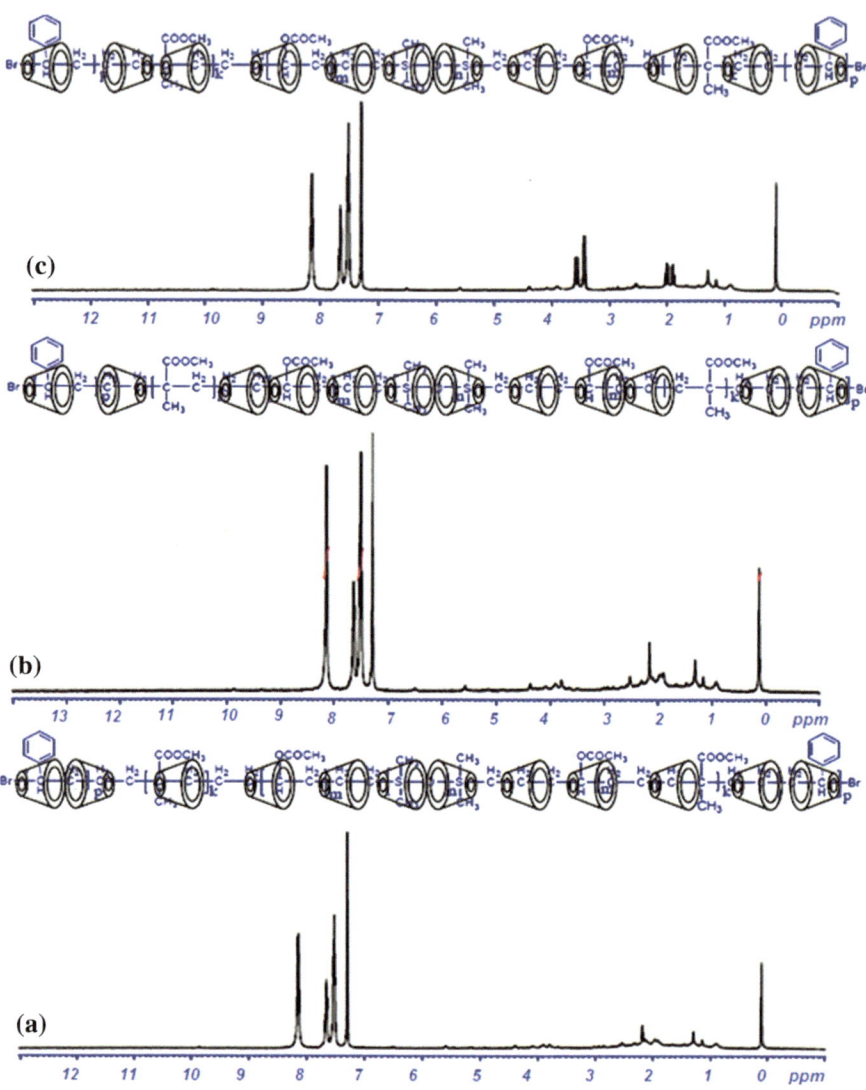

Fig. 5.6 ¹H-NMR of thermoreversible polyrotaxane of PSt-b-PMMA-b-PVAc-b-PDMS-b-PVAc-b-PMMA-b-PSt block copolymers at **a** 22 °C, **b** 55 °C and **c** 22 °C

Figure 5.7 shows the XRD spectra obtained for the PSt-b-PMMA-b-PVAc-b-PDMS-b-PVAc-b-PMMA-b-PSt at 22 °C (Fig. 5.7a), temperature increased to 55 °C (Fig. 5.7b) and temperature decreased from 55 to 22 °C (Fig. 5.7c). Characteristic peaks of PDMS occurring at $2\theta = 8, 13, 18$, characteristic peaks of PVAc occurring at $2\theta = 13$, characteristic peaks of PMMA occurring at $2\theta = 16\text{-}18$ and characteristic peaks of PSt occurring at $2\theta = 14, 12, 11, 8, 6\text{–}17, 25\text{–}9$. At 22 °C, crystalinity percent of PVAc and PMMA is 11 and 28 % respectively (Fig. 5.7a).

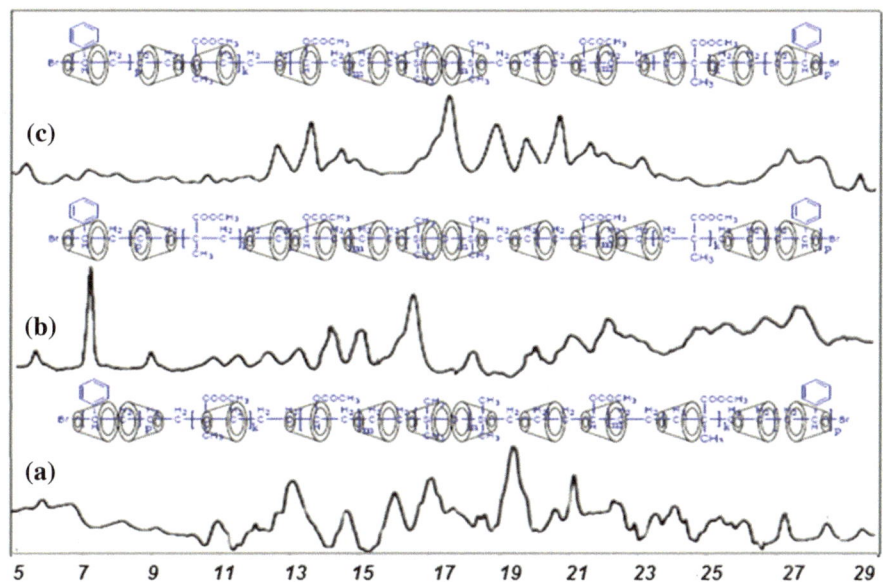

Fig. 5.7 XRD of thermoreversible polyrotaxane of PSt-b-PMMA-b-PVAc-b-PDMS-b-PVAc-b-PMMA-b-PSt block copolymers at **a** 22 °C, **b** 55 °C and **c** 22 °C

By increasing temperature from 22 to 55 °C, crystalinity percent of PVAc and PMMA becomes 24 and 16 % respectively (Fig. 5.7b), which indicated migration of CD from PMMA to PVAc. By decreasing temperature from 55 to 22 °C, crystalinity percent of PVAc and PMMA becomes 13 and 26 % respectively (Fig. 5.7c), which indicated migration of CD from PVAc to PMMA [1, 2, 9].

Figure 5.8 shows the ^{1}H-NMR spectra obtained for the PSt-b-PMA-b-PVAc-b-PDMS-b-PVAc-b-PMA-b-PSt at 22 °C (Fig. 5.8a), temperature increased to 55 °C (Fig. 5.8b) and temperature decreased from 55 to 22 °C (Fig. 5.8c). The integral of the peaks associated with protons adjacent to the styrene and vinyl acetate (6.5–7 ppm and 1.2–2.1 ppm respectively) were observed to increase with increasing temperature from 22 to 55 °C which integral of the peaks associated with protons adjacent to the methyl acrylate (3.5–3.6 ppm) were observed to decrease, relative to the integral of the protons in the PMDS (Fig. 5.8b). This reflects the mobility of the γ-cyclodextrin in block copolymer structure, upon decreasing temperature to 22 °C, integral of the peaks associated with protons adjacent to the styrene and vinyl acetate the copolymers undergo decrease and integral of the peaks associated with protons adjacent to the methyl acrylate undergo an increase to reach the initial integral (Fig. 5.8c) [1, 2, 9].

Figure 5.9 shows the XRD spectra obtained for the PSt-b-PMA-b-PVAc-b-PDMS-b-PVAc-b-PMA-b-PSt at 22 °C (Fig. 5.9a), temperature increased to 55 °C (Fig. 5.9b) and temperature decreased from 55 to 22 °C (Fig. 5.9c). Characteristic peaks of PDMS occurring at 2θ = 8, 13, 18, characteristic peaks of PVAc occurring

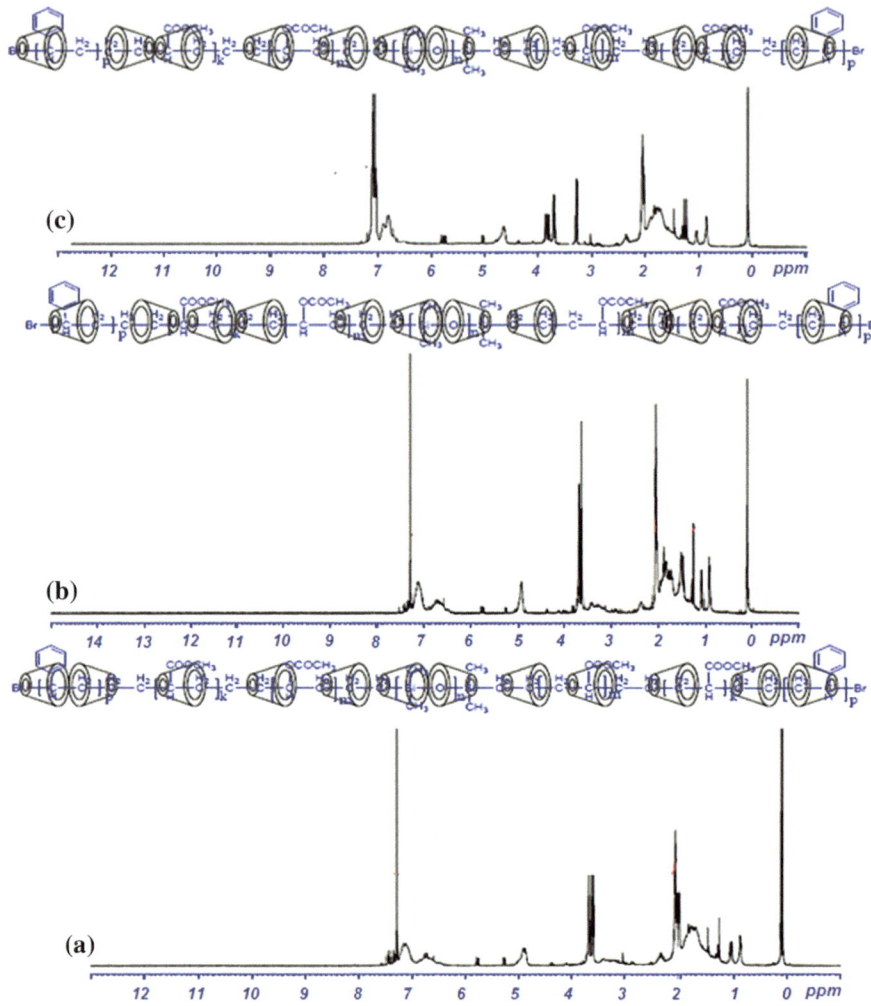

Fig. 5.8 ¹H-NMR of thermoreversible polyrotaxane of PSt-b-PMA-b-PVAc-b-PDMS-b-PVAc-b-PMA-b-PSt block copolymers at **a** 22 °C, **b** 55 °C and **c** 22 °C

at 2θ = 13, characteristic peaks of PMA occurring at 2θ = 21–23 and characteristic peaks of PSt occurring at 2θ = 14, 12, 11, 8, 6-17, 25-9. At 22 °C, crystalinity percent of PVAc and PSt is 18 and 16 % respectively (Fig. 5.9a). By increasing temperature from 22 to 55 °C, crystalinity percent of PVAc and PSt becomes 6 and 10 % respectively (Fig. 5.9b), which indicated migration of CD from PVAc and PSt to PMA. By decreasing temperature from 55 to 22 °C, crystalinity percent of PVAc and PSt becomes 16 and 11 % respectively (Fig. 5.9c), which indicated migration of CD from PMA to PSt and PVAc [1, 2, 9].

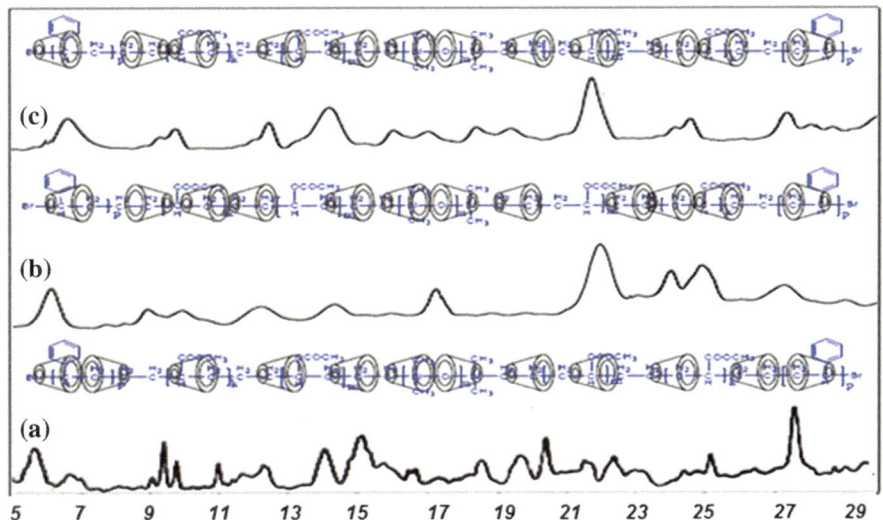

Fig. 5.9 XRD of thermoreversible polyrotaxane of PSt-b-PMA-b-PVAc-b-PDMS-b-PVAc-b-PMA-b-PSt block copolymers at **a** 22 °C, **b** 55 °C and **c** 22 °C

5.4 Summary

The unique temperature-dependent polyrotaxans were successfully synthesized via ATRP of St and PVAc initiated with Br-PDMS-Br/γ-CD using CuCl/PMDETA as a catalyst system at 60 °C. The results in this study strongly supported our assumption that the control of temperature can induce the block-selective inclusion complexation to form the polyrotaxanes, where γ-CD could be threaded exclusively on the middle PDMS block. These polyrotaxanes are of major interest because of their possibility to form amphiphilic pentablock copolymers of PSt-b-PVAc-b-PDMS-b-PVAc-b-PSt by hydrolysis of the PVAc blocks. ^{1}H-NMR technique was utilized to analyze microstructure of the pentablock copolymers as well as the number of PSt, Br-PDMS-Br/γ-CD and PVAc units in the block copolymer. In second method, the temperature-dependent polyrotaxans were successfully synthesized via ATRP of MA, MMA, St and PVAc using CuCl/PMDETA as a catalyst system at 60 °C and Br-PDMS-Br as macroinitiator. The results in this study strongly supported our assumption that the control of temperature can induce the block-selective inclusion complexation to form the polyrotaxanes, where γ-CD could be threaded exclusively on the middle PDMS block. These polyrotaxanes are of major interest because of their possibility to form amphiphilic block copolymers of PSt-b-PMMA-b-PVAc-b-PDMS-b-PVAc-b-PMMA-b-PSt and PSt-b-PMA-b-PVAc-b-PDMS-b-PVAc-b-PMA-b-PSt by hydrolysis of the PVAc blocks. ^{1}H-NMR technique was utilized to analyze microstructure of the block copolymers as well as the number of PSt, PMA, PMMA, Br-PDMS-Br and PVAc units in the block copolymer.

5.5 Concluding Remarks

Poly(dimethyl siloxane) (PDMS)-based triblock and pentablock copolymers were successfully synthesized using atom transfer radical polymerization (ATRP) of vinyl acetate (VAc), styrene (St), methyl acrylate (MA) and methyl methacrylate (MMA) monomers at 60 °C using CuCl/N,N,N′,N″,N″-pentamethyldiethylenetriamine (PMDETA) as a catalyst system initiated from boromoalkyl-terminated poly (dimethyl siloxane) (Br-PDMS-Br) macroinitiator. There was a very good agreement between the number-average molecular weight calculated from ^{1}H-NMR spectra and that calculated theoretically indicating the living/controlled characteristic of the reaction. The novel synthesis of poly(phenylene oxide) (PPO) copolymers with various monomers introduced a controlled radical polymerization process that allows the use of 4-bromo-2,6-dimethyl phenol (BDMP) as an aromatic ether nucleus in a number of new vinyl monomers synthesis like VAc, St and MMA. The monomodal distribution and very narrow molecular weight distribution with high stability further confirmed the controlled nature of the radical copolymerization of BDMP. Recently, attention has been paid on inclusion complex (IC) formed by cyclodextrins (CDs) and inorganic polymers which offer different sites of binding and may be selectively threaded by CDs. The first observation that CDs have formed a complex with inorganic polymers at room temperature without sonic energy is reported in this study. These kinds of complexes may provide a new way of creating new organic-inorganic hybrids and other functional supramolecular architectures, especially inclusion complexes of γ-CD and OH-PDMS-OH or Br-PDMS-Br.

After synthesis and characterization of PDMS based block copolymers, the unique temperature-dependent polyrotaxanes were successfully synthesized via cobalt mediated radical polymerization of various monomers. The results in this study strongly supported our assumption that the control of temperature can induce the block-selective inclusion complexation to form the polyrotaxanes, where γ-CD could be threaded exclusively on the middle PDMS block. ^{1}H-NMR and XRD techniques were utilized to analyze microstructure of the block copolymers as well as the number of PSt, PMA, PMMA, Br-PDMS-Br and PVAc units in the block copolymer.

References

1. Semsarzadeh MA, Amiri S (2013) J Inorg Organomet Polym 23:553–559
2. Semsarzadeh MA, Amiri S (2013) J Inorg Organomet Polym 23:432–438
3. Semsarzadeh MA, Amiri S (2013) Bull Mater Sci 36:989–996
4. Yu-Cai W, Ling-Yan T, Yang L, Jun W (2009) Biomacromolecules 10:66–73
5. Giancarlo M, Marco D, Vittorio C (2008) J Polym Sci A Poly Chem 46:4830–4842
6. Neeraj K, Majeti NVR, Domba AJ (2001) Adv Drug Deliv Rev 53:23–44
7. Dai XH, Dong CM, Yan DY, Wei Y (2006) Biomacromolecules 7:3527–3533

8. Fujita H, Ooya T, Yui N (1999) Macromol Chem Phys 200:706–713
9. Semsarzadeh MA, Amiri S (2013) J Incl Phenom Macrocycl Chem, 77:489–499. doi: 10.1007/s10847-013-0330-1
10. Harada A, Kamachi M (1990) J Chem Soc Chem Commun 19:1322–1323
11. Harada A, Kamachi M (1992) Nature 356:325–327
12. Harada A, Kamachi M (1993) Macromolecules 26:5698–5703
13. Harada A, Kamachi M (1993) Nature 364:516–518
14. Harada A, Kamachi M (1994) J Am Chem Soc 116:3192–3196
15. Harada A, Kamachi M (1995) Macromolecules 28:8406–8411
16. Semsarzadeh MA, Amiri S (2013) Progress in silicones and silicone-modified materials, 8:103–110
17. Semsarzadeh MA, Amiri S (2012) Silicon 4:151–156
18. Braunecker WA, Matyjaszewski K (2007) Prog Polym Sci 32:93–146
20. Matyjaszewski K (2005) Prog Polym Sci 30:858–875
20. Debuigne A, Caille JR, Jerome R (2005) Macromolecules 38:5452–5458
21. Huang F, Gibson WH (2005) Prog Polym Sci 30:982–1018
22. Joachim S, Markus H, Axel HEM, Helmut R (2000) Macromol Rapid Commun 21:1342–1346